K's Manual for

Organic Laboratory

Hongbum Kim

Dongguk University

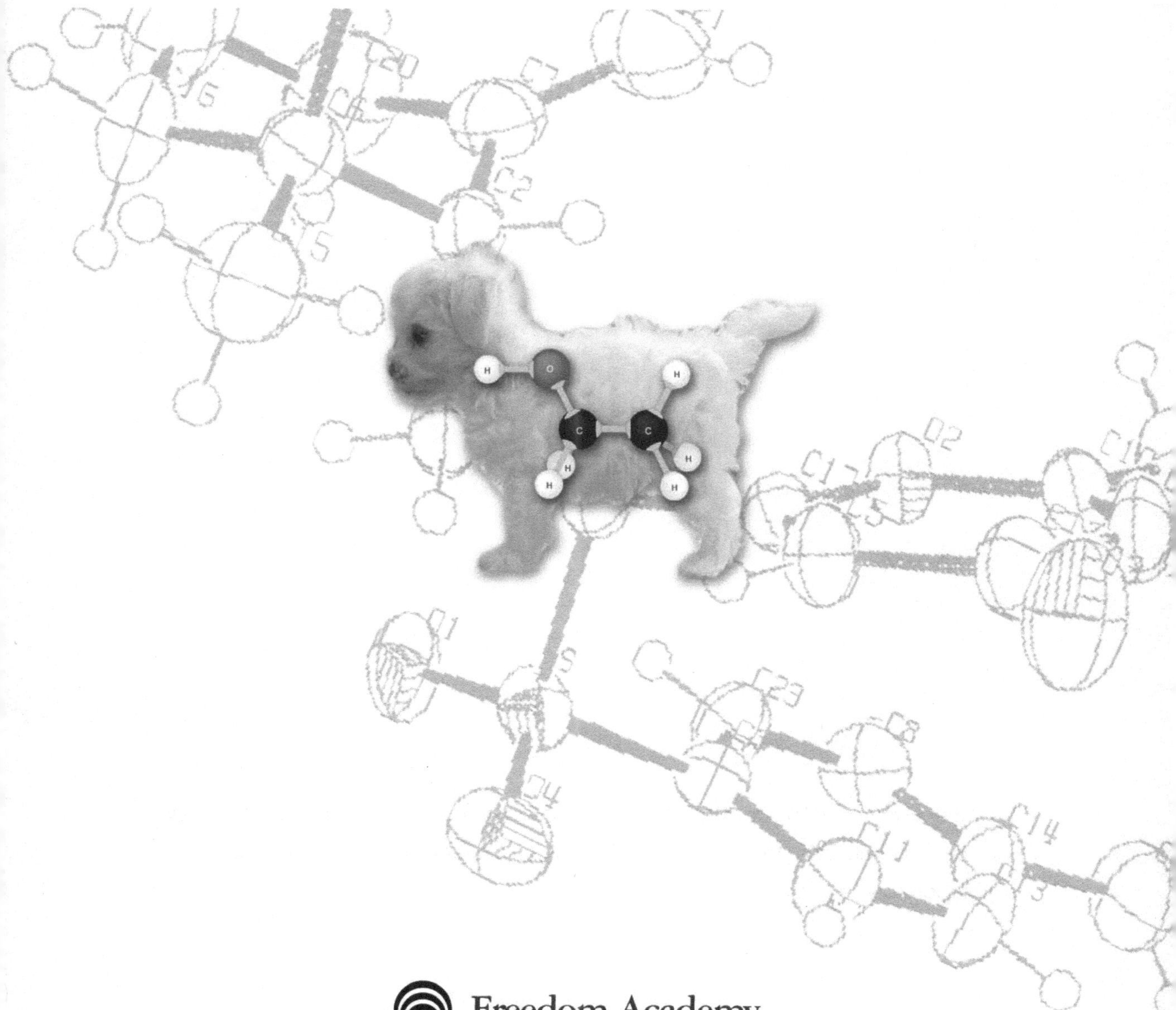

Freedom Academy

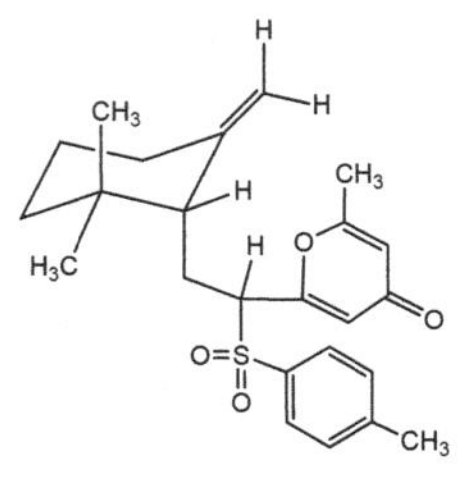

The X-ray crystallography structure in the background of the cover front page is the key intermediate for total synthesis of Forskolin, which was synthesized in our laboratory in 2001.

Preface

This book is a laboratory manual for organic chemistry, which is rewritten in a new format after examining the previously published organic chemistry laboratory textbooks, taking only the good points and enhancing some deficient parts.

First of all, my principal goal is to present the organic structural theories based on the fundamental techniques and the experimental results obtained from the organic laboratory, so that the students can take an interest in organic chemistry. Furthermore, the ultimate aim is to encourage the students to do their own research on organic chemistry based on these newly found interests.

This book will give the students an opportunity to contact with the experimental procedures, results and compounds by carrying out the experiments. They will be able to understand the theoretical parts more easily after becoming familiar with this basic knowledge from experimental data and results. They can also check the validity of the organic chemistry knowledge by experiencing the experimental procedures and results with which they are acquainted in the laboratory. Many existing organic laboratory textbooks have a large amount of contents overlapping with those of the textbooks for organic chemistry lecture. However, the introduction of each chapter in this book does not contain the general contents that are already provided in organic chemistry textbooks. This book aims at giving the students a practical and intermediate knowledge for organic compounds and organic reactions that they will meet during their course of studying organic chemistry.

This book mainly consists of five parts: the **assent** for safety in the laboratory, the **relationship between the physical property and the structure of organic compounds**, the **organic reactions,** the **appendix** and the **index**. The **assent** mentions the safety cautions that the students must take in the laboratory. Recently, organic compounds are misconceived as the main cause of the environmental pollution and as being extremely dangerous. In fact, organic compounds can be very safe and useful when the appropriate safety measures are taken. In order to put the emphasis on this fact, the assent is placed in the beginning. The **relationship between the physical property and the structure of organic compounds** chapter offers the students a chance to meet the common organic solvents and various simple organic compounds. Students can learn the fundamental experimental techniques required for the manipulation of organic reactions such as **dissolution, Thin Layer Chromatography, distillation, extraction** and **recrystallization.** It also emphasizes on the basic knowledge of organic chemistry for finding out the physical property and the reactivity of organic compounds.

The **organic reactions** chapter mainly deals with the reactions that are used in the conversion of basic functional groups and attempts to explain the method of each experiment in related to the **reaction mechanism**. There are many cases which are impossible to get the expected results using the existing organic chemistry laboratory textbooks. Also, an insufficient explanation for the reaction condition delays the experiment time. As a result, the students might lose their interests and confidence in interests organic chemistry rather than enjoying the experiments and feeling rewarded. They might even have a prejudice that organic experiments are too difficult. To overcome this situation, this book checks every detail of each experimental procedure – the amount of reagents, reaction condition, monitoring of the reaction – so that the reaction result can be **reproducible** and the procedure manipulation can be carried out efficiently.

The **appendix** and the **index** provide supplementary information for each chapter.

I would like to acknowledge the perspiration and the endeavor of many preceding authors of organic laboratory textbooks. I thank the undergraduates in the organic division of department of chemistry at Dongguk University for participating in the experiments. I deeply thank Yoon–Sun Jang, Soon–Ghi Won, and Wan–Jung Kim for their contribution in preparing the manuscript, and Sun–Hwa Jeong for her assist in translating this book into English. Without their genuine enthusiasm and spontaneous participation, this book would not have been made.

I express my gratitude to Prof. P. A. Grieco and Prof. Dr. W. Wiegrebe who made me feel competent and confident as an organic chemist. Prof. Sung–Hoon Kim, Prof. Young–Hwan Lee, Prof. In–Hyeong Yeo, Prof. Won K. Seok and Prof. Shigenori Kashimura carefully looked over this book and gave priceless encouragement through difficult times.

The last but not the least, I dedicate this book to my father, late Dr. You Sun Kim, who guided me to pursue the career as an organic chemist , to my mother, Ms. Young Hee Kim and my grandmother, late Ms. Cha Sook Lee who have always looked after me with patience and love.

A few words to the students

The laboratory to an organic chemist is just like the surgery to a surgeon, or the field to a football player. No surgeon will think about other things nor talk on the cellular phone during the operation. Nor will the football player during the game. It is the same with the organic chemist. In the laboratory, you should only concentrate on the experiment that you are carrying out at the moment.

You only get what you put into it. Therefore use your experimental time wisely and efficiently.

1. You must wear safety glasses in the laboratory. Read the laboratory safety cautions before carrying out the experiment.

2. This book is intended to let you understand organic chemistry based on the experimental results. Each chapter consists of **introduction**, **experimental procedure**, and **note**. You can immediately carry out the experiment as you read the experimental procedure and the note without any prior knowledge in the **introduction** and the **reaction mechanism**. Then you can write down your observations and results in the book. However, make sure that you have a thorough knowledge of the structures of the compounds being used before entering the laboratory.

3. During every organic experiment, there is a time when you have to wait for the result of the reaction. While you are waiting, **draw out the structures of the compounds and the reaction mechanism** in the blank spaces in your manual as many times as possible. The introduction deals with the general application and more advanced details about the chemical compounds or the reactions used in the experiment. Read the introduction after you have fully understood the reaction procedure and learned the structure of the compounds mentioned.
   After obtaining the experimental results, Refer to your textbook and other advanced organic chemistry books to study more about the theories and the reactions mentioned in the introduction part. There is no time for waiting absent–minded in the organic laboratory but only time to think about the organic chemistry experiments and theories.

4. Suppose you are now carrying out an organic chemistry experiment. If you expect to get the right result at once just after you exactly follow the procedure, which means you do not really understand the nature of organic reactions. Many organic reactions written in the textbooks have gone through hundreds and thousands of trials, failures, and innumerous mistakes. It is more likely that you will not get the stated result – especially when it is your first time – regardless of how well the reaction procedure is devised. Therefore, there is no reason to be disappointed even if you did not get the correct result in your first trial.
   An experiment is not determined by one result only. It is a process of finding a way to get the same result on repetition. Although it may suggest that you are not yet fully experienced, it is perfectly normal to get the wrong result in the first trial.
5. **All the results obtained from the laboratory are important**, whether they are right or wrong. If you can explain how and why you got the wrong result, the failure would give you more valuable information for the next experiment than a successful result by chance would. In order to obtain a satisfactory result, you should not feel bothered to repeat the experiment many times or be afraid that you might fail. You must just keep on challenging.
6. I hope that this book will help many young prospective scientists to enjoy organic chemistry experiments and eventually lead them to the challenging world of organic chemistry, which expands day by day.
7. Instead of just acquiring knowledge and training of organic chemistry, I encourage you to challenge the new, unfamiliar phenomena, to find a way to reproduce them and to make them your own by innovating a better way. Furthermore, bear in mind that you are learning and inheriting the great tradition of science whenever you conduct an experiment.

Suggestions for the laboratory instructor

The contents of most organic laboratory textbooks have not been changed in the last 30 years since I was a undergraduate student in chemistry major. And I understand it is not easy to switch to a new laboratory textbook and to get used to the new experimental contents all over again since there is a great variation on the facility and the level of students depending on the university. However, I believe that many professors of organic chemistry recognize the great need for an alternative that will change the contents and the operation of the organic chemistry experiments. This book is written as one of the alternatives. I hope this book will bring out an active discussion for better organic chemistry education, especially for laboratory.

1. In order to make up for the weakness of the existing laboratory textbooks, which do not give an appropriate method for monitoring the reaction procedure and checking the product, TLC is applied in every experiment in this book. To make checking of TLC easier, UV. active aromatic compounds are used where possible. Moreover, the same organic compounds are used repetitively in the basic techniques and the organic reactions for the full comprehension of the compounds by the students and also for recycling.
2. The role of the teaching assistant (T.A.) is very important in the organic laboratory. In most universities, the teaching assistants are the graduate students. But it is almost impossible for one TA to manage 20–30 students in the laboratory. So, I suggest that you select a number of students from junior and senior classes who have keen interest in organic experiments. And then you train each of them in one specific topic, and let them be the laboratory helpers. If you operate the laboratory with the TA and the laboratory helpers, it will be easier to give the students accurate information about the procedures such as setting up the apparatus.
3. It is good to conduct a various set of experiments; however, I suggest that you give the students a make–up time to repeat the experiments which they got an unsatisfactory result. After 4–5 weeks of individual experiments, you may give them a chance to present their own results and discuss freely with their fellow students in order to activate an ardent atmosphere in the organic chemistry laboratory.
4. In using this book, precede chapter 1 (the relationship between the physical property and the structure of organic compounds) at first, and then deal with the

organic reactions. Having thoroughly studied the basic techniques, the students would be able to conduct the organic reactions without difficulty. Regarding the operation of the basic technique section, I suggest that you divide the students into small groups and assign each group to different topic of the basic technique sections rather than having everyone do the same experiment. At first, conducting five experiments simultaneously may turn out to be tough. However, after 2 or 3 weeks, you will see a significant change in the experimental environment if you keep encouraging the students to explain to each other about the experimental procedures and results. It will then draw out a spontaneous discussion among the students and generate an active and sincere atmosphere in the laboratory. In addition, conducting various experiments simultaneously can result in the increase of efficiency of the equipment. If time permits, apply the same method in the organic reaction chapters.

An example of syllabus for one semester is as the following:

	Week	Experiment Contents
Techniques	1st week	Introduction and Check-in
	2nd ~ 6th	1.1 Dissolution and Polarity 1.2 Thin Layer Chromatography and Isomers 1.3 Distillation and Polarizability 1.4 Extraction and Acid Dissociation Constant (pKa) 1.5 Recrystallization and Melting Point
	7th	Midterm examination and make-up session
Organic reactions	8th~14th	2.1 Esterification 2.4 Oxidation 2.5 Reduction 2.6 Nitration 2.8 Aldol Condensation 3.1 Dye 3.4 Diels-Alder reaction
	15th	Make up session
	16th	Final examination and Check-out

For **modulation** of the laboratory, we arranged the necessary glassware and reagents on the laboratory benches for each experiment topic as you can see below. As a result, the number of glassware is kept to a minimum and the moving line of students for the preparation of the required reagents is shortened. As a result, the experiment time is remarkably reduced. Although the modulation of the laboratory method has already been applied in physics and physical chemistry in which a large number of equipments are used for an experiment, we are the first to confirm that it could also be applied as effectively in organic chemistry.

Computer	Hood for acidic compound	Hood for basic compound	
Extraction	Extraction	Recrystal-lization	Recrystal-lization
TLC	TLC	Dissolution	Dissolution
Distillation	Chemical Balance	Vacuum filter	Distillation
Evaporator	Evaporator		

Module of Chapter 1

Computer	Hood for acidic compound	Hood for basic compound	
Esterification	Esterification	Diels Alder	Reduction
Oxidation	Dye	Nitration	Nitration
Aldol condensation	Chemical Balance	Vacuum filter	Aldol condensation
Evaporator	Evaporator		

Module of Chapter 2 and 3

5. The study guidebook, which contains data on the observations and the results of each experiment to help you evaluate the performance of your students, is in preparation. Contact the publisher if you need it. Also, any inquires about the contents of the book or about operating the laboratory will be gladly answered.

Contents

Chapter 03

Intermediate Organic Reactions

Chapter 04

Appendix

LABORATORY SAFETY

Accident Case 1

Professor K. Barry Sharpless who won the 2001 Nobel Prize in Chemistry for his discovery of “chiral catalysts” wrote this account of an accident. (This case shows how important it is to wear safety glasses in the laboratory, and is frequently quoted whenever there are safety lessons in the United States of America.)

Many of you know that I was blinded in one eye during a lab accident in 1970, shortly after I arrived at MIT as an assistant professor. I always wore safety glasses whenever I was at my bench and, while I felt that I was conscientious about observing safety measures, my experience proves that you can't be too cautious about wearing glasses.

As I prepared to go home from the lab during the early hours of the morning of the accident, I checked in to see what my co–workers were doing and then returned to my bench, removed my safety glasses, and put on my parka. As I was walking to the door, I passed the bay where a first–year graduate student was flame–sealing an NMR tube. I asked how it was going and he replied, "Good. I've got it sealed."

He was sealing off the tube at atmospheric pressure in a big liquid nitrogen bath, a procedure the two of us had discussed though neither had ever performed it. I stopped by his bench, picked up the tube from the bath and held it to the light. The tube immediately frosted over and, as I wiped it to better see the contents, I noticed that the solvent level was exceedingly high. Suddenly the solvent dropped to a normal level and, though I instantly realized condensed oxygen had been sealed in the NMR tube, I was quite literally unable to move a muscle before it exploded. Glass fragments shredded the cornea, penetrated the iris, and caused the partial collapse of one eye. My only other injuries were superficial facial cuts.

My first two weeks at ophthalmic hospital were spent totally immobilized with both eyes bandaged. The pain was terrific, but my fear was even greater: I had been warned that when my eyes were uncovered there was a small chance I might be blind in both eyes due to sympathetic ophthalmia. Because eyes are walled off from the rest of the body in utero, eye protein driven into the blood stream can raise an immune response that leads to the killing of an uninjured eye. My disappointment at having no functional vision in my injured eye was, needless to say, surpassed by my joy at retaining full vision in my good eye.

The lesson to be learned from my experience is straightforward: **there is simply never an adequate excuse for not wearing safety glasses in the laboratory at all times**.

(Cited from Chemworld by KCS vol. 40 no.1 in 2000)

Safety Caution 1

1) **Safety glasses on, contact lenses off**: In order to protect the eyes, safety glasses must be worn. And it is advised not to wear contact lenses.
2) **Move carefully and wear the appropriate shoes**: Be careful when carrying a glassware containing reagents, and do not rush in the lab. It is recommended to wear sneakers, which cover the top of your feet. Do not wear slippers or high–heeled shoes.
3) **Use tongs or gloves**: Hot glass instruments should be handled with tongs. Wear suitable gloves to protect the hands when handling compounds with high reactivity.

Accident Case 2

There was an explosion accident in a laboratory of University of California at Fullerton on April 3, 1999, which called for more than 10 fire engines and 40 firefighters.

Luckily, the fire ended with only a few scratches. This accident happened because alcohol wastes were discarded in the acid wastebasket. As a result of an intense reaction, the wastebasket exploded, and the instruments around it were also broken. For this reason, putting alcohol wastes in glass containers is banned, but this is not kept well by students or researchers. If there were any inflammable organic solvents nearby, there could have been serial explosions, which would have caused a disaster.

Acidic or basic wastes should be kept apart from organic wastes to minimize the confusion and also to prevent possible serial explosions in case of an accident.

(Cited from http://mail.kcsnet.or.kr/chemworld/plan2/plan2_006.htm)

Safety Caution 2

4) **Discard the wastes in a fixed container**: The reagents must be discarded in the provided containers. Acidic reagents or reagents that form peroxides should not be mixed with other substances before being discarded. Especially, the organic solvents should be separated into solvents which include halogen, and solvents which do not before being removed using separate waste tubs. Do not discard strong acids, bases or organic solvents directly into the sink.

5) **On handling reagent bottles**: Use a reagent only after checking the tag of the reagent bottle first. Do not take any reagents, which produce gases out from the hood. The reagent bottle should be held with both hands – one on the body and the other one on the bottom. Do not hold the bottle with one hand by the mouth of the bottle.

6) **Check the stopper**: After taking chemicals from containers, always make sure the stoppers are closed properly.

7) **Do not put back the reagent**: Generally speaking, reagents should not be returned to their containers once they are out. Some reagents can be put back on special occasions, but only when they are not contaminated. The moisture in the air can contaminate a reagent.

8) **Write down information on the container**: All the containers containing chemicals used for a long term should have names on them. The containers of synthesized chemicals or intermediates should have the date of synthesis, weight as well as name and structure on them.

9) **Use dry instruments**: The instruments – such as the spatula that is used to take chemical from a container – should be clean and dry.

10) **On heating**: When heating up a liquid, make sure the mouth of the test tube is not pointed at other students. Inflammable liquids must be heated using a water bath.

11) **Use the hood**: If poisonous or pungent gases are produced during the

experiment, it should be carried out under the hood. Do not place your head under the hood. Chemicals which are carcinogenic, bring tears to your eye or have low vapor pressure (e.g. bromine, fuming sulfuric acid, acetyl chloride) should also be handled under the hood.

12) **On adding water to an acid**: When diluting a concentrated acid, add the acid slowly to the water with stirring. Not the other way round.

Emergency Measure

1) **In case of a fire**: Turn off all the heat sources including the burners and the electricity, move away any inflammable substances, and fight the fire using the chemical fire extinguisher or sand. When an organic solvent, which does not dissolve well in water, catches fire, do not use water to extinguish the fire.
2) **In case of clothes on fire**: Do not rush, and extinguish the fire using a blanket or a laboratory coat. If the fire is not close to the face, the chemical fire extinguisher can also be used. And you may use water as well if the fire was not caused by an organic solvent.
3) **In case of a burn caused by a reagent**: Wash immediately with plenty of clean water. If an acid caused the burn, use diluted sodium hydrogen carbonate; if a basic reagent caused it, wash with diluted acetic acid. And give an appropriate medical treatment afterwards.
4) **In case of getting a reagent in the eye**: After washing immediately with plenty of clean water, seek medical care.
5) **In case of inhaling poisonous gas**: Move to a well-ventilated place immediately and breathe in the fresh air. If a large amount of poisonous gas is taken in, seek medical help immediately.

Websites concerning chemistry and safety

Society for Chemical Hazard Communication (http://www.schc.org/)
National Safety Council (http://www.nsc.org)
MSDS (http://www.msdssearch.com)
Korean Chemical Society (http://www.kcsnet.or.kr/kcs/environment/environ_index.htm)

Check-In Glassware and Apparatus I

Group :　　　　Name :

<table>
<tr><th colspan="2">Glassware or Apparatus / Quantity</th><th>Necessary quantity</th><th>Actual quantity</th><th>Comment</th></tr>
<tr><td rowspan="3">Distillation set</td><td>Distillation Head</td><td>1</td><td></td><td></td></tr>
<tr><td>Condenser</td><td>1</td><td></td><td></td></tr>
<tr><td>Adapter</td><td>1</td><td></td><td></td></tr>
<tr><td>Reflux condenser</td><td>14/20</td><td>1</td><td></td><td></td></tr>
<tr><td>Separatory funnel</td><td>100mL or 250mL</td><td>2</td><td></td><td></td></tr>
<tr><td>Round bottom flask</td><td>100mL</td><td>2</td><td></td><td></td></tr>
<tr><td>Two-neck flask</td><td>100mL</td><td>1</td><td></td><td></td></tr>
<tr><td>Filter flask</td><td>250mL</td><td>1</td><td></td><td></td></tr>
<tr><td>Graduated cylinder</td><td>100mL</td><td>1</td><td></td><td></td></tr>
<tr><td rowspan="3">Erlenmeyer flask</td><td>25mL</td><td>3</td><td></td><td></td></tr>
<tr><td>50mL</td><td>3</td><td></td><td></td></tr>
<tr><td>100mL</td><td>3</td><td></td><td></td></tr>
<tr><td rowspan="3">Beaker</td><td>50mL</td><td>2</td><td></td><td></td></tr>
<tr><td>100mL</td><td>2</td><td></td><td></td></tr>
<tr><td>250mL</td><td>2</td><td></td><td></td></tr>
<tr><td colspan="2">TLC chamber</td><td>1</td><td></td><td></td></tr>
<tr><td colspan="2">Büchner funnel</td><td>1</td><td></td><td></td></tr>
<tr><td colspan="2">Stopper</td><td>1</td><td></td><td></td></tr>
<tr><td colspan="2">Watch glass</td><td>1</td><td></td><td></td></tr>
<tr><td colspan="2">Conical funnel</td><td>1</td><td></td><td></td></tr>
</table>

Check In Glassware and Apparatus II

Glassware or Apparatus \ Quantity	Necessary quantity	Actual quantity	Comment
Magnetic stirrer	1		
Transformer	1		
Heating mentle	1		
Lab jack	2		
Utility clamp	2		
Clamp holder	2		
Ring	2		
Ring holder	2		
Cork ring	1		
Forcep	1		
Spin-bar	1		
Spatular	1		
Glass rod	1		
Thermometer	1		

Date:

Student:

Laboratory Instructor:

Assent for Safety in the Laboratory

Read the safety cautions in the organic laboratory carefully, and write down the five main points and the reasons that you agree to keep all the time in the laboratory.

It should be confirmed and co-signed by the laboratory instructor.

1.

2.

3.

4.

5.

Date :

Student Name and Signature

Laboratory Instructor Signature

CHAPTER

01

The relationship between the physical property and the structure of organic compounds

1.1 Dissolution and Polarity

Keywords : hydrophilic, hydrophobic, aromaticity, polar, nonpolar, isomer

When the table salt is added in water and stirred thoroughly, it disappears and the water maintains its transparency. However, if more than a certain amount of the salt is added, it no longer dissolves and stays in its solid form at the bottom of the container.

Why can we not see the salt when it is dissolved in water?

What is the difference between the solid and the dissolved salt?

When two liquid substances are mixed, either they cannot be distinguished from each other or they do not mix to form two distinctive layers. For example, ethanol cannot be distinguished when it is added to water. But when cooking oil is added to water, two separated layers are formed. It can be defined that ethanol dissolves in water while cooking oil does not.

Then, why does ethanol dissolve in water, when oil does not ?

It is interesting that the origin of the word "**dissolve**" not only means "mix with a liquid and become part of it" but also "**disappear**." This shows us that people have already astutely observed and described this phenomenon from a long time ago.

In chemistry, when the bulk solid dissolves in the liquid, it is described as '**soluble**'. But when it does not dissolve well, it is described as '**insoluble**'. For instance, when a solid is soluble in a certain liquid, it cannot be seen in our eyes since the coagulated solid is divided into tiny bits and spread out evenly into the liquid. In this case, the liquid takes the role of a knife or a grinder that chops the solid into small pieces.

When the liquids are mixed and cannot be distinguished from each other, they are described as '**miscible**'. In contrast, when they form separated layers, they are described as '**immiscible**'. In the case of miscible liquids, two different molecules corresponding to each liquid spread out evenly into the same space.

The similarity of soluble and miscible phenomena is that the mutual contact surface of each compound is increased remarkably and the movements between the two substances are free.

These properties are one of the crucial factors for a successful chemical reaction which heavily depends on the exchange of energy between the reactants. When the reactants −either solid or liquid− do not mix well together, they have less contact surface and cannot exchange the activation energy needed for the chemical transformation.

In the view of organic chemists, matters in nature are divided into two groups: one is the water–like substance (**hydrophilic** or **lipophobic**) and the other is the oil–like substance (**hydrophobic** or **lipophilic**). One important characteristic is that **they do not mix with each other**. If you look around, you will soon realize how efficiently we utilize it in our daily life –in cooking, cleaning, cosmetics, etc–. In fact, our body is made up of 70 % of the water–like substances and 30 % of the oil–like substances. Both of them are essential for our body but they have different functions. Roughly speaking, organic chemists are defined as those who study properties and reactivities of the oil–like substance.

Now we approach a fundamental, yet important question about matter in nature:

Why do water and oil not mix well together?

In organic chemistry, these two types of substance are described as **polar** (water–like or hydrophilic) and **nonpolar** (oil–like or hydrophobic) depending on their **molecular structures.**

The characteristics of the molecular structure depend on the distance and the angle of the chemical bonds, which connect atoms to build a molecule. Based on the tremendous amount of data on the organic compounds which have been accumulated for over 100 years, a number of elaborated theories about chemical bonds have been extensively developed by the enthusiastic organic chemists. As a result, much of properties and reactivities of organic substance can be understood and interpreted by the molecular structure. Furthermore, the relationship between bioactivity and the molecular structure is currently under active investigation.

Understanding the relationship between the physical property and the molecular structure is the most important step in learning modern organic chemistry. This chapter is devoted to explain the relationship between the solubility of organic compounds and their molecular structures.

Apparatus

Experiment I

⇨ 10 mL test tube, test tube stand, 1.00 mL Micro pipette, Vortex mixer

— It is recommended to use Micro pipette for measuring the solvent and Vortex mixer for shaking if possible —

Experimental procedure

1. Place 2.0 mL of the following compounds in 10 mL test tubes and mix each of them well with the equal amount of a solvent.
2. Check whether any layers are formed, and mark either "miscible" or "immiscible"
3. Interpret the results based on the structure.

Name	Structural formula	H_2O	Hexane
H_2O			
Hexane			
Carbon tetrachloride			

Note I

1. In general, when **3.0 g** of the substance is dissolved in **100 mL** of solvent at 25℃ and the solid cannot be seen with the naked eye, the substance is said to be **'soluble'** in the corresponding solvent. However, in this textbook, 50.0 mg of the solid sample is dissolved in 3.0 mL of solvent in order **to observe the difference in solubility behavior only**. In the case of a liquid sample, an equal volume of the liquid sample and the solvent are mixed for the same reason.

2. The **intermolecular force** determines the physical property – e.g. solubility, boiling point, melting point, etc – of the substance. If there is no intermolecular force, no liquid or solid can be formed and every single substance will exist as a gas. The major intermolecular forces are the **dipole–dipole interaction** for **polar compounds**, **van der Waals interaction** for **nonpolar compounds**, and the well–known **hydrogen bonding**. The dipole–dipole interaction will be introduced in this section. The van der Waals interaction will be covered later in **1.3 Distillation**.

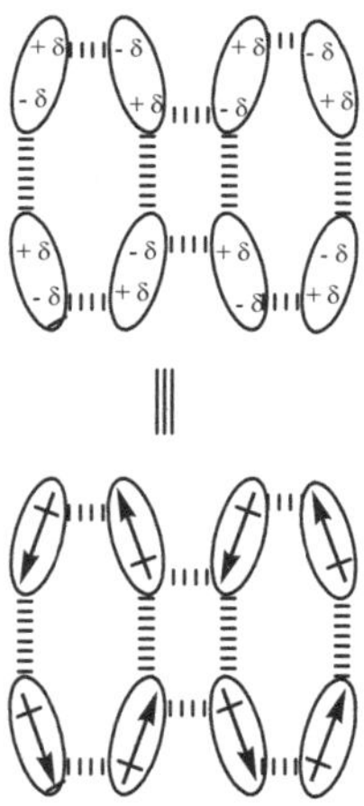

3. A polar compound has a partial charge separation in its electron cloud and holds a **permanent dipole** which is measured by the dipole moment μ (Debye). For example, hydrochloric acid (HCl, μ =1.11D), water (H_2O, μ=1.84D), methanol (CH_3OH, μ=2.87D) and acetone ($(CH_3)_2CO$, μ=2.69D) are polar compounds. In contrast, a nonpolar compound has no charge separation and a permanent dipole moment that is close to zero Debye. For example, hydrogen gas (H_2, μ=0D), oxygen (O_2, μ=0D), hexane (C_6H_{14}, μ=0.09D), and carbon tetrachloride (CCl_4, μ=0D) are nonpolar compounds. The dipole moment is derived from the magnitude of the partial charges and the distance separating them.
The partial (+δ) charge is facing the partial (–δ) charge of the polar compounds. The resulting electrostatic attraction force results in the energy stabilization. The amount of this energy is about 1~3 kcal/mol. It is smaller than that of the ionic compound, but large enough to give out a considerable amount of heat when it is condensed into a liquid.

In conclusion, polar compounds have a relatively large intermolecular force based on the permanent dipole.

4. When the polar liquid substances are mixed, the same electrostatic attraction occurs between the dipoles of each substance resulting in their uniform distribution in the solution. This interaction of the permanent dipoles in different molecules is called the **dipole–dipole interaction.**

   The polar and nonpolar substances do not mix together since there is very weak stabilizing interaction between them. Therefore, they maintain their own association state, which gives rise to the distinctive separated layers when they are mixed. (**Likes dissolve likes**)

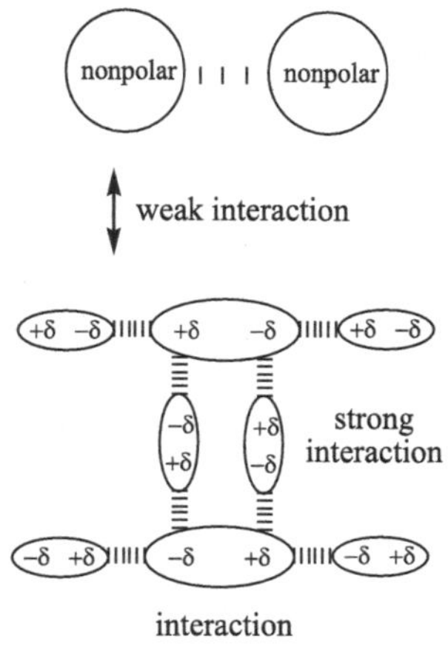

5. The permanent dipole of a polar compound is the vector summation of the **individual polar bonds** and the **lone–pair electrons** in the molecule. A polar bond arises from the different **electronegativities (EN)** of two atoms participating in the bond. The greater the difference in electronegativity between the bonded atoms, the higher the polarity of the bond. For example, the electronegativities of carbon and halogen —fluoride, chloride— or oxygen are large enough that carbon–halogen or carbon–oxygen bond has a considerable polarity. In contrast, carbon–carbon and carbon–hydrogen bonds are not polar since their electronegativities are very close. In general, a bond which has difference in electronegativity between 0.7 and 2.0 is called the polar bond. Meanwhile, the electronegativity of a nonpolar bond is smaller than 0.7 and that of an ionic bond is larger than 2.0 respectively.

element	EN
F	4.0
O	3.5
Cl	3.0
N	3.0
C	2.5
H	2.1
Si	1.8

   Water (H_2O, 1.84D) consists of two oxygen (EN=3.5)—hydrogen (EN=2.1) bonds and two lone–pair electrons on the oxygen atom. The lone–pair electrons on the oxygen atom are far apart from the nucleus to increase the distance between the partial charges in addition to the large charge separation and contribute to the large dipole moment of water.

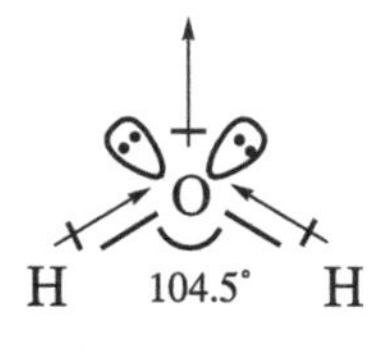

$\varepsilon = 78.5$ (at 25℃)

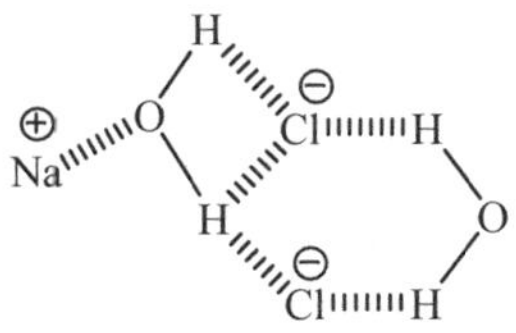

6. The hydroxyl group can readily form a hydrogen bond with other compounds, especially with water and results in high solubility in water. Ionic compounds such as salt dissolve very well in water because of the strong hydration of both positive and negative ions by the water molecule.

7. Hexane (C_6H_{14}) is a nonpolar compound since it is made up of single bonds between carbon (EN=2.5) and hydrogen (EN=2.1). Compounds made up of hydrogens and carbons are called **hydrocarbon compounds.** They play an important role of providing skeletons for the organic substances. Nonpolar compounds are not soluble in water since they do not carry a permanent dipole required for the interaction. In the case of nonpolar molecules, they hold **the induced dipole moment** in contrast to the permanent dipole moment in the polar compounds. Induced dipole moments are caused by **van der Waals interaction** that will be dealt thoroughly in **1.3 Distillation.**

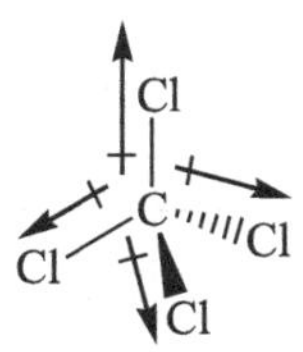

μ=0 D

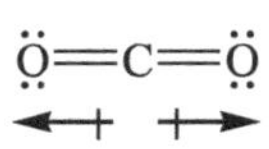

μ= 0D

8. Not all the molecules with polar bonds are polar. The C–Cl bond in carbon tetrachloride is polar due to the large difference in electronegativity between carbon and chloride atom. However, the equivalent individual bond moments cancel out each other because of the tetrahedral structure of carbon. Therefore the carbon tetrachloride becomes a nonpolar compound. The same principle is applied to carbon dioxide.
In contrast, chloroform, methylene chloride and carbon monoxide have the permanent dipole moments.

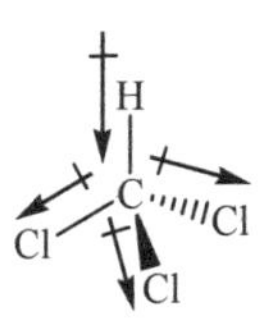

μ=1.89 D

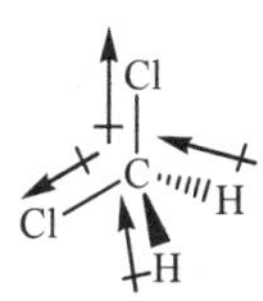

μ=1.60 D

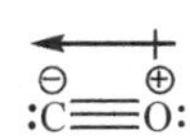

μ=0.10 D

9. In conclusion, compounds that are soluble in water are called **polar** and compounds that are not soluble in water are called **nonplar** based on the "**Likes dissolve Likes**" principle.

[Part A]

Experiment 2

Apparatus

⇨ 10 mL test tube, test tube stand, 1.00 mL Micro pipette, Vortex mixer

Experimental procedure

1. Place 2.0 mL of the following compounds in 10 mL test tubes and mix each of them with the equal amount of a solvent.
2. Check whether the layers divide, and mark either "miscible" or "immiscible"
3. Interpret the results based on the structure.

Name	Structural formula	H_2O	Hexane
Methanol			
Propanol			
Pentanol			
Phenol			
Ethylene glycol			

Note 2

$H-\ddot{O}-H$

$R-\ddot{O}-H$

1. Alcohol is a compound containing a hydroxyl group (–OH) bonded to a sp^3–hybridized carbon. It is a derivative of water in which one of the hydrogen in water is replaced by a various hydrocarbon including an aromatic hydrocarbon. Methanol (wood alcohol) which can cause blindness is the simplest alcohol. Ethanol (grain alcohol) is widely used in alcoholic beverages and fuel additives in gasohol. Both of them are important organic solvents and valuable starting materials for the organic laboratory and the industry.

2. The presence of the hydroxyl group greatly influences the physical and chemical properties of alcohols. The polarity of the oxygen–hydrogen bond and the ability to form hydrogen bonds with other strong electronegative atoms—fluorine, oxygen, nitrogen—result in the high solubility in polar solvents and the high boiling point.

3. The OH group in the alcohol is called the **hydrophilic group** – like water – and the R group is the **hydrophobic group** – dislike water or like hexane. These effects compete with each other to decide the solubility characteristic of the corresponding alcohol. For example, methanol is soluble in water but insoluble in hexane since the hydrophilic effect caused by OH group overrides the hydrophobic effect by —CH_3. But 1–pentanol does not mix well with water because of the large hydrophobic effect by the large alkyl group. Instead, it dissolves well in the nonpolar solvent like hexane. In the case of propanol, it dissolves either in water or in hexane since its hydrophilic effect and hydrophobic effect are complementary and work in either solvent. It can be concluded that even if a molecule has a hydroxyl group, its solubility behavior depends on the characteristic of the R group.

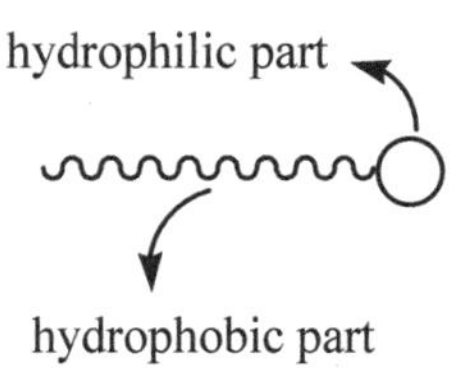

4. A surfactant molecule contains a hydrophilic group and a hydrophobic group. Soaps and detergents are the typical surfactants.

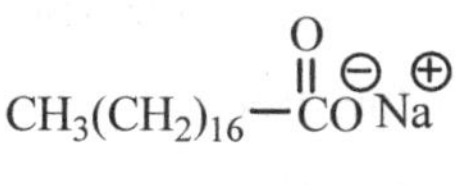

sodium stearate (soap)

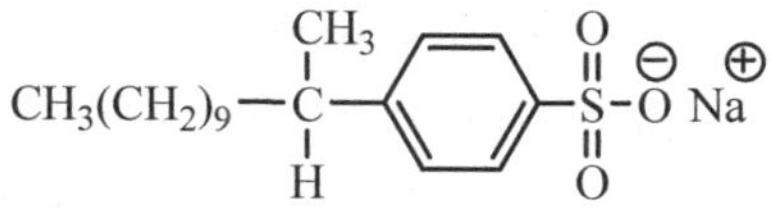

linear alkyl sulfonate (detergent)

5. Phenol contains a benzene ring. Benzene which has a zero dipole moment is one of the typical nonpolar organic compounds and shows a strong hydrophobic property. The structure of benzene is quite different from that of the simple unsaturated compounds. All of its carbon–carbon bonds have the same length, which is the intermediate between the typical single and double bonds. It does not undergo the addition of halogen and the reduction by hydrogen gas which are characteristic reactions of the double bond in alkene. The remarkable stability of benzene is called **aromaticity**. Benzene and its derivatives are called the **aromatic compounds** that are substantially stabilized by the π electron delocalization.
   For the complete electron delocalization, the molecule should meet the following structural requirements.
   a. The molecule must be **cyclic, conjugated** and **planar**.
   b. Each atom of the molecule must have a p orbital which is perpendicular to the plane of the ring
   c. The molecule must have **(4n+2) electrons—Hückel Rule**

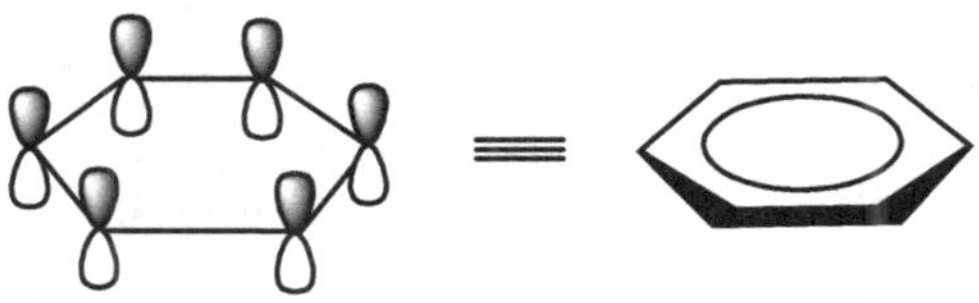

Conversely, monocyclic and nonplanar molecules with 4n electrons are especially unstable and called **antiaromatic**. For example, cyclobutadiene with 4π electrons is antiaromatic and highly unstable.

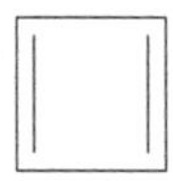

cyclobutadiene (4π electrons) unstable

Nowadays, the aromaticity of organic compounds is determined by observing the unusual chemical shifts of hydrogen bonded to the outer periphery of the ring. Hydrogens in benzene are highly deshielded by the ring current effect caused by the delocalization of πelectrons. Its absorption occurs far downfield at 7.24. This effect

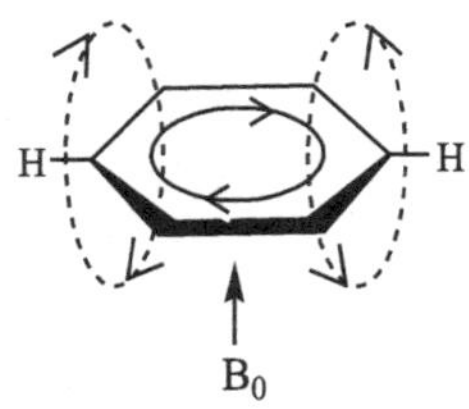

external magnetic field

is called the **diamagnetic anisotropy** or the **diamagnetic shielding**. (It may be dealt later after you learn about the aromatic compounds and/or the NMR spectroscopy of the organic substance)

6. The **arene** is used to describe the aromatic hydrocarbons. Benzene is the parent arene. Just as the group derived by the removal of an H from an alkane is called an alky group (R–), the group derived by removal of an H from an arene is called an **aryl group** (Ar–)

7. The **annulene** is a cyclic hydrocarbon with continuous alternation of single and double bonds. Cyclobutadiene and benzene are named [4] annulene and [6] annulene, respectively. Both [14] annulene and [18] annulene were synthesized to confirm the Hückel Rule for aromaticity. They were found to have aromaticity by observing the large differences in the chemical shift of the outer and inner hydrogens.

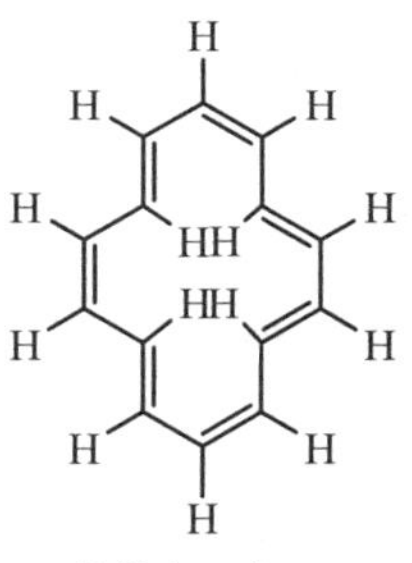

[14] Annulene
(aromatic, n=3)

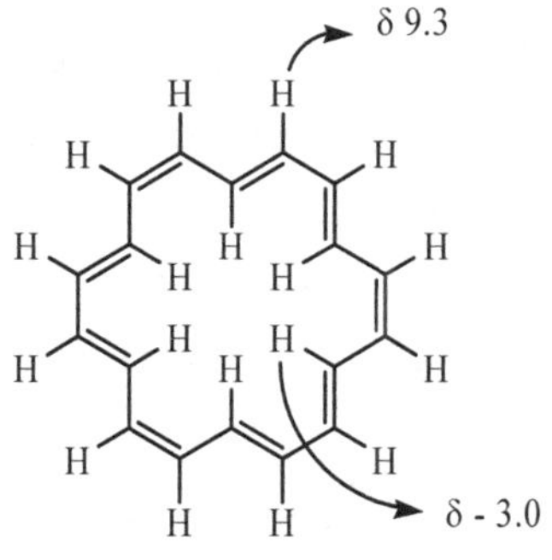

[18] Annulene
(aromatic, n=4)

[18] annulene – 60℃, ^{1}H NMR, protons outside the ring δ 9.3
protons inside the ring δ –3.0

8. Some alcohols have more than one hydroxyl group. One of them is ethylene glycol, which is used as the automobile antifreeze. Ethylene glycol is soluble in water but not in hexane since it has a high dipole moment and two hydroxyl groups, which cause the strong hydrophilic effect.

[Part B]

Apparatus

⇨ 10 mL test tube, test tube stand, 1.00 mL Micro pipette, Vortex mixer

Experimental procedure

1. Place 3.0 mL of the following compounds in 10 mL test tubes and mix each of them with 1.0 mL of a solvent.
2. Check whether layers divide, and mark either "miscible" or "immiscible"
3. Interpret the results based on the structures.

	Structural formula	H_2O	Hexane
1-Butanol			
Isobutyl alcohol			
2-Butanol			
tert-Butyl alcohol			

Note 3

1. As expected from the result of the previous section, 1–butanol is soluble in hexane but not in water. On the contrary, *tert*–butyl alcohol is soluble in water. Both 1–butanol and *tert*–butyl alcohol have the same molecular formula ($C_4H_{10}O$) but the solubility in water is quite different. These are **isomers**. In most organic textbooks, isomers are defined as compounds which have the identical molecular formula but different structural formulae. But in this textbook, isomers are defined as **compounds that have the identical molecular formula but different physical properties** since isomers were first discovered by experimental facts and then interpreted by the molecular structural theory which was developed later. Understanding isomerism in organic substances has been the major driving force in developing the structural theory in organic chemistry—hybridization, resonance, conformation, etc—(refer to **Appendix Table of Isomers**)
   The interpretation of experimental results based on the structural theory is strongly emphasized throughout this textbook for a thorough understanding of modern organic chemistry.

2. Since the liner geometry has more surface area to contact with each other than does the spherical shape, the hydrophobic property is increased as the surface area increases and *vice versa*. As a result, the spherical R–group in *tert*–butyl alcohol does not have enough hydrophobic property to overcome the hydrophilic property caused by the hydroxyl group. As a result, *tert*–butanol is soluble both in water and in hexane.

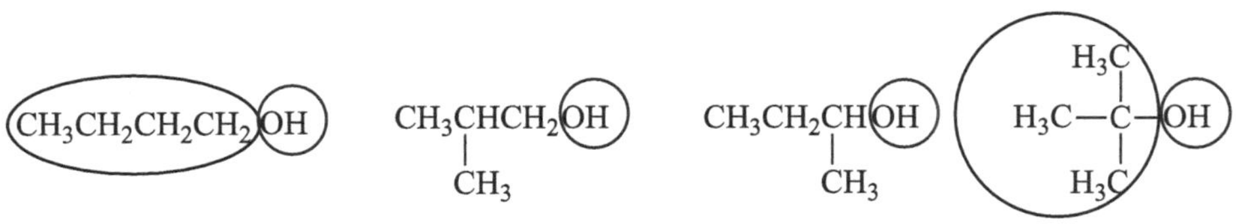

3. The hydrophobic property is closely related to **polarizability** which is a measure of how much the electrons of a molecule are distorted. It is affected by the size and the shape of a molecule and will be dealt in detail in **1.3 Distillation**.

Experiment 3

Apparatus

⇨ 10 mL Test tube, Test tube stand, 1.00 mL micro pipette, Vortex mixer

Experimental procedure

1. Place 2.0 mL of the following compounds in 10 mL test tubes and mix each of them with the equal amount of a solvent.
2. Check whether layers divide, and mark either "soluble" or "insoluble"
3. In the test tubes which are miscible from the procedure 1, add 0.2 g of solid sodium chloride (NaCl) twice and stir thoroughly.
4. Observe if layers divide.
5. Interpret the results based on the structures.

Name	Structural formula	H_2O	Hexane	NaCl
Ethanol				
Diethyl ether				
THF				
Furan				
Dioxane				

Note 4

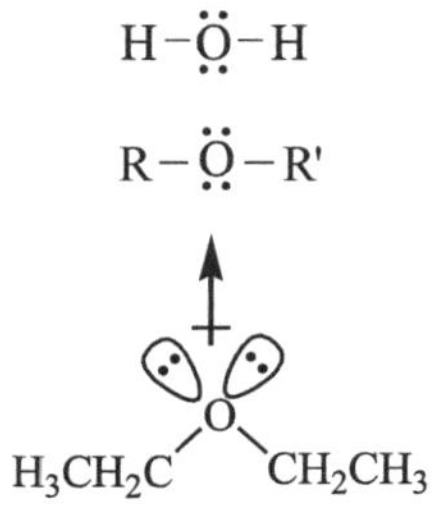

μ = 1.15D

b.p. : 34.6℃

μ = 1.75D

b.p. : 65℃

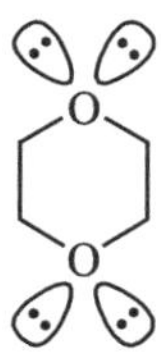

μ = 0 D

b.p. : 100~102℃

μ = 0.66D

b.p. : 32℃
(at 758mmHg)

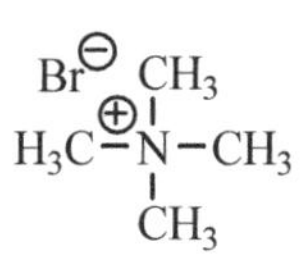

tetramethyl
ammonium bromide

1. Ether is an analogue of water in which both hydrogens are replaced by hydrocarbons or aromatic hydrocarbons. Since they do not have a polar oxygen–hydrogen bond, they have much lower boiling points than the alcohols of comparable molecular weight. For example, the boiling point of diethyl ether is 34.6℃ compared to 117℃ of 1–butanol.

2. Tetrahydrofuran (THF) and dioxane are cyclic ethers. Because of their cyclic structure, they have lone–pair electrons that are more accessible for hydrogen bonding with water than those of diethyl ether. As a result, they are soluble in water. Since the cyclic structure also attributes to an increase in their dipole moments, the cyclic compounds have significantly higher boiling points than the corresponding acyclic compounds.

3. Furan is an aromatic compound since it satisfies the requirements for aromaticity by the contribution of lone–pair electrons on oxygen. Please Refer to the (4n+2) **Hückel rule** in the previous section.

4. When a sufficient amount of salt is added to the miscible mixture of tetrahydrofuran and water followed by a thorough mixing, the water layer and the tetrahydrofuran layer are separated. This is called the **salting–out effect**. The solubility of organic substances in water is markedly reduced by the presence of dissolved inorganic salts especially sodium chloride. For example, tetrahydrofuran and dioxane are completely miscible in water but only slightly soluble in saturated sodium chloride solution (**brine**). The salts which have ions of small radii and concentrated charges are more effective.

5. On the contrary, the enhanced solubility in a salt solution is observed in the ions of large radii and diffused charges. (**Salting–in effect**) It has been reported that benzene is about 40 percent more soluble in 1.0 M aqueous tetramethylammonium bromide than in pure water.

Apparatus

Experiment 4

⇨ 10 mL test tube, test tube stand, 1.00 mL Micro pipette, Vortex mixer

Experimental procedure

1. Place 2.0 mL of the following compounds in 10 mL test tubes and mix each of them well with the equal amount of a solvent.
2. Check whether layers divide, and mark either “miscible” or “immiscible”
3. Interpret the results based on the structures.

Name	Structural formula	H_2O	Hexane
Acetaldehyde			
Benzaldehyde			
Acetone			
Cyclohexanone			
Acetophenone			

Note 5

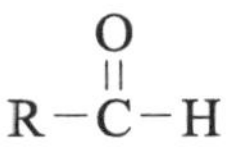

aldehyde

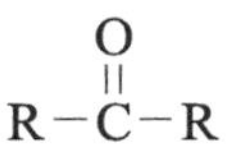

ketone

1. Aldehyde and ketone have a carbonyl group bonded to the hydrogen and the carbon atom. The carbonyl group containing the carbon–oxygen double bond determines most of the physical and chemical properties of aldehydes and ketones. The carbonyl group has a large permanent dipole moment because of the large electronegativity difference between oxygen and carbon in addition to the resonance contribution of the π double bond. This polarity of the carbonyl group makes the low–molecular–weight aldehydes and ketones soluble in water. Carbonyl compounds dissolve in water as long as the hydrocarbon portions of the molecule do not contain more than four or five carbon atoms.

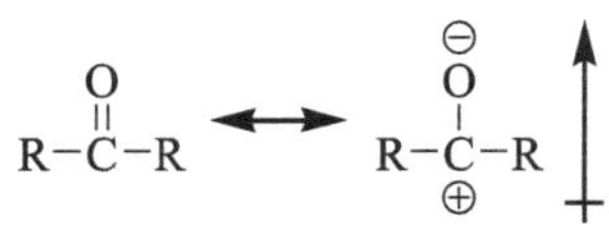

2. Acetone dissolves both in water and in hexane since its hydrophilic group (carbonyl group) and hydrophobic group (alkyl group) give similar contribution for solubility and do not override each other. The same behavior is observed in ethyl alcohol and *tert*–butyl alcohol in the previous section.

3. For more information about cyclic and aromatic compounds, please Refer to the previous section about alcohol and ether.

4. A carbonyl compound may exist in two forms called **tautomers.** (***keto*** and ***enol***) Most simple carbonyl compounds exist primarily in their keto forms: however, 2,4–pentanedione exits as 80% enol since it is stabilized by internal hydrogen bonding.

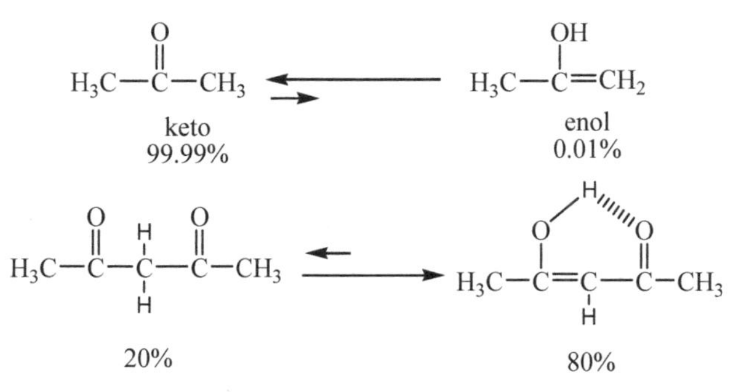

[Part A]

Apparatus

⇨ 10 mL test tube, test tube stand, small spatula balance, 1.00 mL Micro pipette, Vortex mixer,

Experimental procedure

1. Place 50 mg of the solid compound in a 10 mL test tube, and add 3.0 mL of a solvent in each of them. If it turns into a clear solvent, mark “soluble”, and if not, mark “insoluble”.
2. When the compound is liquid, place 2.0 mL of the compound in a 10 mL test tube and add 2.0 mL of a solvent. Check whether layers divide, and mark either “miscible” or “immiscible”.
3. In the test tubes which are immiscible from the procedure 1, 2, add 1.00 M of sodium hydroxide solution (NaOH) drop by drop to observe the change. (Be careful, it is acid/base reaction)
4. Interpret the results based on the structures.

Name	Structural formula	H_2O	Methylene chloride	1.0M NaOH
Acetic acid				
Stearic acid				
Benzoic acid				

Note 6

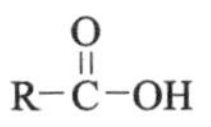

carboxylate

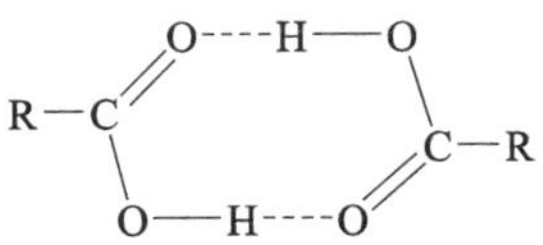

Dimer form of carboxylic compound

1. Carboxylic acid has a carboxyl group which consists of a carbonyl group and a hydroxyl group. The interaction between the carbonyl group and the hydroxyl group affects the physical and chemical properties of a carboxylic acid. The hydrogen of the hydroxyl group in the carboxylic acid is easily removed as a proton by the large resonance contribution in its carboxylate anion than that in alcohol. They are much stronger organic acids than the alcohols.
   In the solid, liquid and even gas phases, the carboxylic acids exist largely in a dimeric form caused by the hydrogen bonding between the carbonyl group and the hydroxyl group. This strong hydrogen bonding influences the boiling points of the carboxylic acids, which are much higher than those of the corresponding alcohols. For example, b.p. of acetic acid is 118℃ and ethyl alcohol is 78.3℃.

2. As expected from the previous discussion about the hydrophilic group and the hydrophobic group, acetic acid is soluble in water whereas stearic acid is insoluble in water.

3. Benzoic acid is a white solid and the simplest aromatic carboxylic acid. Benzoic acid sparingly dissolves in water or hexane since it has a strong tendency to aggregate within itself to form a dimer.

4. The carboxylic acid readily reacts with basic sodium hydroxide to form a salt and water.

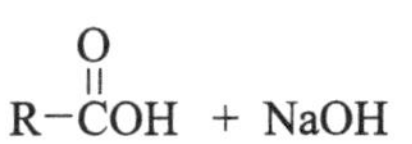

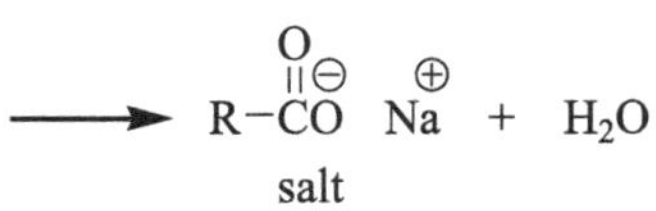

The salt which has a carboxylate anion and a sodium cation is the typical ionic compound. It is very soluble in water. However, the salt of stearic acid is insoluble in the basic sodium hydroxide solution since the hydrophobic property caused by a large R group overrides the hydrophilic property by the ionic bond between carboxylate anion and sodium cation.

[Part B]

Apparatus

⇨ 10 mL test tube, test tube stand, 1.00 mL Micro pipette, Vortex mixer

Experimental procedure

1. Place 2.0 mL of the following compounds in 10 mL test tubes, and mix each of them well with the equal amount of a solvent.
2. Check whether layers divide, and mark either "miscible" or "immiscible"
3. Interpret the results based on the structures.

Name	Structural formula	H_2O	Hexane
Acetic acid			
Ethyl acetate			
Acetamide			
Acetonitrile			

Note 7

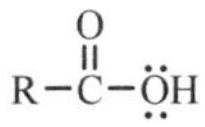
carboxylic acid

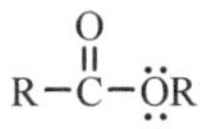
ester

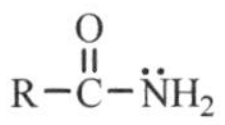
amide

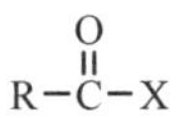
acyl halide

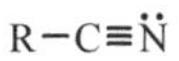
nitrile

1. Ester, amide and nitrile are derivatives of the carboxylic acid. Acid derivatives are analogues of the carboxylic acid in which the hydroxyl group is replaced by another atom or a group. —chlorine for acyl chloride, alcohol for ester and amine for amide— Their reaction with water yields the corresponding carboxylic acids.

2. Ethyl acetate, which has no acidic proton, serves only as a hydrogen–bond acceptor on its oxygen atom in water and shows low solubility in water. (8.6 g per 100 g water) However, acetamide (97.5 g per 100 g water) is very soluble in water since the hydrogen atoms on the nitrogen atom of an amide participate strongly in the hydrogen bonding with water. The strong resonance contribution of an amide will be discussed in **1.2 Thin Layer Chromatography.**

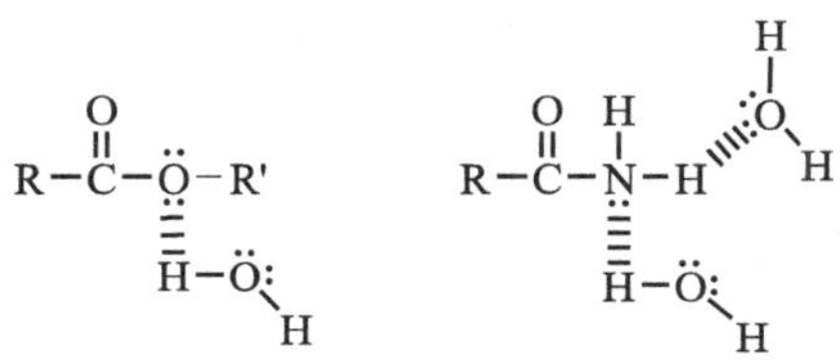

3. Acetonitrile is soluble in water since it has a strong dipole moment caused by the strong electronegativity of nitrogen as well as the lone – pair electrons on the nitrogen at the end of the molecule.

4. When the two compounds have the equivalent electronic distribution but only differ in their nuclei, they are said to be **isoelectronic**. Acetonitrile and methylacetylene are isoelectronic. Their chemical properties are qualitatively similar.

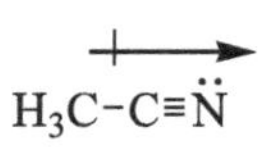
μ =3.44 D

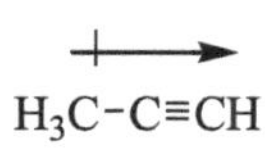
μ =0.784 D

[Part C]

Apparatus

⇨ 10 mL test tube, test tube stand, small spatular balance, 1.00 mL Micro pipette, Vortex mixer,

Experimental procedure

1. Place 1.0 mL of the following compounds in 10 mL test tubes and add 2.0 mL of a solvent in each of them.
2. Check whether layers divide, and mark either "miscible" or "immiscible"
3. In the test tubes which are immiscible from the procedure 1, 2, add 6.00 M of hydrochloric acid (HCl) drop by drop to observe the change. (Be careful, it is an acid/base reaction)
4. Interpret the results based on the structures.

Name	Structural formula	H_2O	Methylene chloride	6.0M HCl
Hexylamine				
Octylamine				
Aniline				
Pyridine				

Note 8

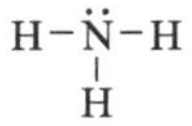
ammonia

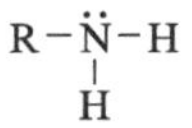
primary amine

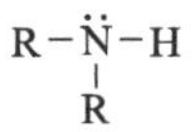
secondary amine

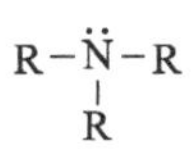
tertiary amine

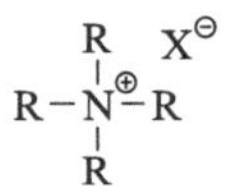
quarternary ammonium salt

1. Amine is an analogue of ammonia in which one or more hydrogens are replaced by hydrocarbons or aromatic hydrocarbons. The most important properties of amines are their basicity and nucleophilicity caused by the lone–pair electrons. The nitrogen–hydrogen bond is less polar than the oxygen–hydrogen bond and the hydrogen bonding in amine is weaker than in alcohol since the electronegativity of nitrogen is smaller than that of oxygen. As a result, the boiling points of amines are lower than those of alcohols of similar molecular weight. For example, the b.p. of butylamine is 78℃ and butyl alcohol 118℃.

2. As shown in the previous result on the solubility of butyl alcohols, 1–butyl alcohol is not soluble in water. But hexylamine is soluble in water since the lone pair electrons in a nitrogen atom are more available for hydrogen bonding with water than those in a oxygen atom of alcohols.

3. The amines readily react with the hydrochloric acid to form quaternary ammonium salts, which are ionic compounds. They are soluble in water. The insoluble amine compounds can be dissolved in an acidic aqueous solution by their salt formation.

$$R\ddot{N}H_2 + HCl \longrightarrow R\overset{\oplus}{N}H_3\overset{\ominus}{Cl}$$

4. The **dipole moment** and the **dielectric constant** are closely related but are not always proportional to each other since the dipole moment is the **microscopic property** (molecular property) and the dielectric constant is the **macroscopic property** (bulk property). However the dielectric constant is useful for making a division of solvents into two broad categories — the **polar solvent** ($\varepsilon > \sim 15$) and the **nonpolar solvent** ($\varepsilon < \sim 15$).
The electrostatic attraction force of the ions or the polar substances becomes smaller when they are dissolved in a solvent with a high dielectric constant. Therefore, the ions or the polar substances have a greater tendency to dissociate in the polar solvent than in the

nonpolar solvent.

It is also useful to classify the polar solvents into two categories: a **protic solvent** that contains an oxygen–hydrogen bond or a nitrogen–hydrogen bond and an **aprotic solvent**. The protic solvents strongly solvate the anions by hydrogen bonding whereas the aprotic solvents do not. It has been reported that the reactivity of the strongly solvated anion in a protic solvent is reduced by the factor of three to four powers of ten in S_N2 substitution or E_2 elimination reaction.

Solvent	Dipole Moment (D)	Dielectric Constant (ε)[a]	Boiling Point (℃)
Polar aprotic			
Acetone	2.69	20.7	56.2
Acetonitrile	3.44	37.5	81.6
DMF	3.86	36.7	153
DMSO	3.9	46.7	189
Polar protic			
Ethanol	1.66	24.6	78.3
Methanol	2.87	32.7	64.7
Water	1.84	78.5	100
Nonpolar aprotic			
n-Hexane	0.085	1.88	68.7
Carbon tetrachloride	0[b]	2.24	76.7
Diethyl ether	1.15 [b]	4.34	34.6
Chloroform	1.75	4.80	61.1
THF	1.75	7.58	65.0
Ethyl acetate	1.66	6.08	77.1
Methylene chloride	1.01	9.10	39.8
Nonpolar protic			
t-Butanol	1.66	12.5	82.4
Cyclohexanol	1.86	15.0	161

a. T = 25℃ *b.* T = 20℃

Keywords for Dissolution

Summary for Polarity

1.2 Thin Layer Chromatography and Isomers

Keywords : partition law, monitoring, co−spot, R_f, isomer

Chromatography is an analytical method used in all parts of science. The Russian botanist, Mikhail Tswett invented chromatography at the turn of last century. He separated many chlorophyll and carotenoid pigments such as carotin and xanthophyll by passing the mixture solution through a glass tube filled with finely ground calcium carbonate. Separated chemical species would appear as colored vertical lines in the tube. This is why it was named chromatography. (Chroma means 'color' and graphein means 'record' in Greek.) Afterwards, the applications of chromatography have been developed explosively and 12 scientists won the Nobel Prize on their research related to chromatography.

Thin Layer Chromatography (**TLC**) is a solid–liquid absorption chromatography made by applying silica gel or alumina powder on a glass, plastic or aluminum plate. Based on the **partition law**, chromatography accompanies the partition of materials between the **stationary phase** and the **mobile phase**. In TLC, the stationary phase (solid) is normally the silica gel layer or alumina and the mobile phase is the developing solvent.

The analyzing material travels on the TLC plate by the capillary phenomenon. It is selectively absorbed by the intermolecular interaction between the silica gel and the developing solvent. In general, the stationary phase (silica gel or alumina) is more polar than the developing solvent. As a result, the polar compounds are more easily bound to the silica gel and move slower when the developing solvent moves. They appear at the lower part of the TLC plate. In contrast, when the material is nonpolar, it is less bound to the silica gel and appears at the upper part since it travels far along with the developing solvent. (refer to **Likes dissolve Likes principle**)

The order of bonding tendency with silica gel is:

Acids and bases > amides > alcohols > aldehydes and ketones > halides > esters > unsaturated hydrocarbons > saturated hydrocarbons

The significance of TLC is explained below

1) Before carrying out an organic reaction, the **purity of the reagents** including the solvents must be confirmed. Only one spot should appear on TLC if the substance is pure. When it is contaminated, several spots will be observed which usually equal the number of the components. If any impurities were included in the reaction, it would not proceed smoothly because of the possible side reactions or the formation of by–products. Especially, the reagents that have been opened for a long time or borrowed from another laboratory could have been decomposed. Therefore, the purity of the starting materials must be checked before carrying out the experiment.

2) As the reaction proceeds, **monitoring**–checking the decrease of the starting material and the formation of a new compound–becomes important to adjust the reaction conditions. Since TLC takes only 2~3 minutes and needs less than a milligram of the reaction mixture, it is a simple and fast way to monitor the reaction without influencing the whole percent yield.

3) Many methods – NMR, IR, or MS – are available for identifying the resulting product but they require expensive instrumentation and excessive time. In TLC, the comparison of $\mathbf{R_f}$ values of the spots between the product and the authentic compound might provide the identity of the compounds.

Throughout this textbook, the use of thin layer chromatography is strongly emphasized since it provides valuable information to the organic researchers.

In this experiment, we will find the relationship between $\mathbf{R_f}$ and the **structure of organic compounds**. The compounds selected here will be used in the later section of this book. It is advisable that you become familiar with these compounds and their R_f values beforehand.

Experiment I

Apparatus

⇨ TLC plate, hot plate, UV lamp, forcep, pencil, graduated ruler

Glassware

⇨ TLC chamber, 10 mL vial, capillary tube, spoid, pipette

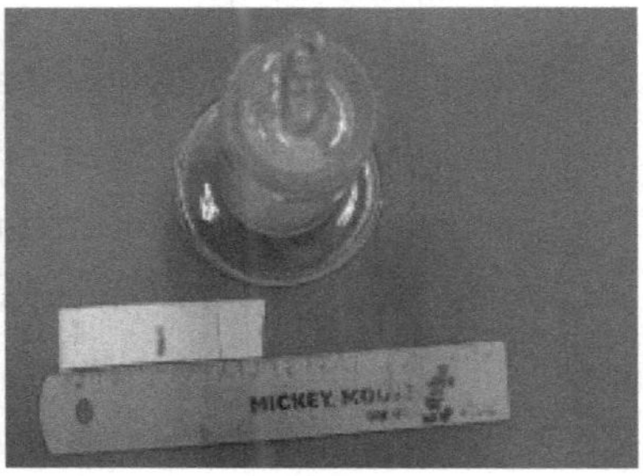

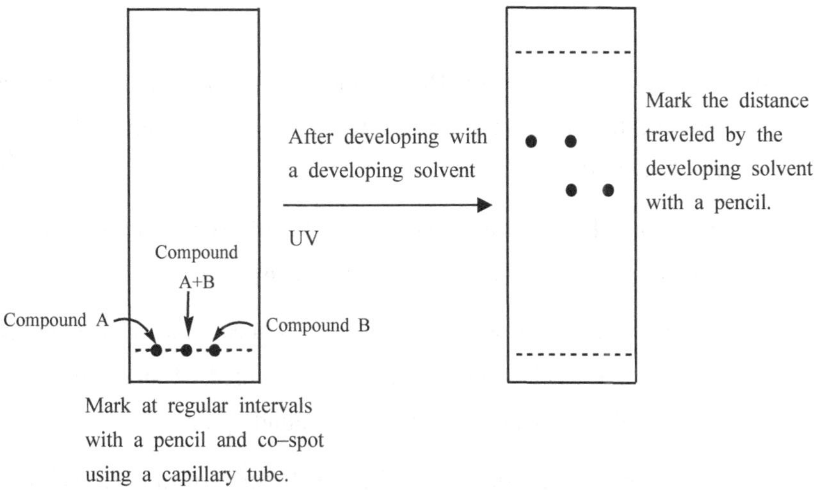

Experimental procedure

Fix the samples with a specific compound, change the ratio of the developing solvent.

1. Preparation of the sample — put 50.0 mg of each sample in separate vials, and add 3.0 mL of ethyl acetate.
   a) benzaldehyde b) benzyl alcohol
2. Preparation of the developing solvent — Make the developing solvents in various ratios of ethyl acetate and hexane. [Note 1]
   Make developing solvents in ratios of
   ethyl acetate : hexane = 8:1, 4:1, 1:1, 1:4, 1:8
   Pour the developing solvent to the TLC chamber about 0.3 cm of from the bottom.
3. Use a pencil to lightly draw a line across the plate about 0.6 cm from the bottom. Then mark three spots on this line which will indicate the position of the samples. [Note 2]
   The first mark is spotted with the first sample and the third with the second sample. The second mark is spotted with the first and the second sample together (**co–spot**). Put the resulting TLC plate in the TLC chamber standing on the end using forceps. [Note 3 or 4]
4. Develop the TLC plate. Before the solvent front reaches 0.5 cm from the top, remove the plate from the chamber and allow it to dry. [Note 5]
5. Illuminate the resulting plate with the ultraviolet light. Encircle each spot lightly with a pencil. [Note 6]
6. With the use of a ruler, measure the distances that the developing solvent and each compound have traveled. Using this data, calculate the R_f value of each compound. [Note 7]
7. Write down the conditions of the developing solvent and draw the resulting TLC plate. [Note 8, 9,10,11]

Date of experiment : **Room temperature :**

Observation :

Name	Structural formula	R_f (E.A. : Hex.)				
		8 : 1	4 : 1	1 : 1	1 : 4	1 : 8
Benzaldehyde						
Benzyl alcohol						

R_f* : Write down the condition of the developing solvent when tracing the TLC plate.

〈Tracing the TLC plate〉

Checked by ______________________

Note

1. Close the lid to prevent the developing solvent from volatilizing so the ratio of the developing solvent does not change.

2. The ballpoint pen should not be used since the ink from the pen may be dissolved into the developing solvent causing an undesirable situation.

3. Use the capillary tube when putting spots on the TLC plate. Contact the end of the capillary tube with the sample, make the sample come up through and then put a spot on the TLC plate. Used capillary tubes should be washed immediately with acetone before being used again. Dip the end of the capillary tube in acetone and let the acetone suck in and out. Repeat it two or three times.

4. **Co–spotting** is a very useful technique in comparing the samples, which have similar developing patterns on TLC. In addition, co–spotting is also useful when comparing the reaction product with the authentic compound. If the co–spot of two compounds appears as one under the condition of more than two different developing solvent systems, these compounds are very likely to be the same compound.

5. As the developing solvent (or the mobile phase) ascends the plate, the spotted sample slowly moves up along with the plate on the coated silica gel (or the stationary phase). This process usually takes several minutes.

6. Most organic compounds are colorless. Therefore, a certain procedure is required to detect these compounds on TLC. This process is called **visualization**. The three most common methods are by using UV lamp (ultraviolet lamp), iodine, or dyeing reagent.
   1) Visualizing using UV light.
   A fluorescent material made of zinc silicate or cadmium sulfide is coated on the TLC plate and this material gives out a fluorescent light under the UV light. When the TLC plate is exposed to the ultraviolet light, the organic substance quenches

the background fluorescence and appears as a black spot. Two kinds of wavelengths are available (254 nm and 365 nm). The experiments in this book use 254 nm.

2) The iodine test uses the characteristic of iodine to stain organic compounds with a brown or yellow spot. When you put a small iodine crystal in a big–mouthed bottle with a screw lid, the bottle is filled with iodine vapor. Then you place the developed TLC plate inside the bottle, and the spots will be seen on the TLC plate.

3) In the method of using a dyeing reagent, dip the developed TLC plate in the prepared dyeing solution. Then take it out, heat it on the preheated hot plate and check the colored spots. (See **Appendix for preparation of dyeing reagent**)

7. The distance a particular compound travels up a TLC plate depends on the intermolecular forces of the sample between the stationary phase and the mobile phase. Typical intermolecular interactions include the **dipole–dipole** and **van der Waals** interactions as well as the **hydrogen bonding**. In principle, the more polar the compound on the TLC plate, the stronger it binds the stationary phase and the shorter it travels.

8. When the spot of the sample does not appear under the visualization methods mentioned above, it means that the sample is too dilute. In that case, mark it several times on the same spot with a capillary tube. Wait until the spot is dry before putting another spot. The alternative way you could use is to increase the concentration of the sample.

   When the sample on the TLC plate after the visualization appears as a long strip, either the concentration of the sample is too high or there are many compounds in the sample. If the concentration is too high, dilute the concentration of the sample and carry out the experiment again. If many samples are standing in a line, decide whether the sample is contaminated.

9. The **R_f (ratio of front or retention factor)** value indicates how far the sample is developed, so

$$R_f = \frac{\text{the distance of the sample developed}}{\text{the distance of the developing solvent developed}}$$

By definition R_f values will vary from 0.0 to 1.0

The R_f value may vary depending on the solvent used. If the polarity of the developing solvent becomes higher, the sample is less absorbed to the silica gel and more soluble in the developing solvent. As a result, it travels farther.

Under the same solvent system, the R_f value is constant for any given compound and closely related to the structure of that compound. Based on this principle, the comparison of R_f values may be used to tentatively identify an unknown compound.

The R_f values are sometimes very close. Putting a **co–spot** rather than comparing each R_f value separately enables you to detect every small difference. It should be noted that different compounds may have the same R_f values so we recommend not making any hasty decision on the basis of TLC alone.

10. When the ratio or the polarity of the developing solvent is fixed, we can observe the developing tendency based on the structures of different samples. When the polarity of the sample increases, the resulting R_f becomes smaller since it is more absorbed to the silica gel and less soluble in the developing solvent. In order to develop the sample farther, the polarity of the developing solvent should also be increased as discussed before.

11. The comparison of the R_f value of a spot with that of an authentic compound under the different developing solvent systems is a simple but reliable method of choice to confirm the identity of a compound. But as mentioned above, do not make any hasty decision on the basis of TLC alone.

Experimental procedure

Experiment 2-1

1. Preparation of the sample — put 50.0 mg of each sample in separate vials, and add 3.0 mL of ethyl acetate.
   a) benzyl alcohol b) benzyl bromide
2. Preparation of the developing solvent — Make the developing solvent in a ratio of ethyl acetate : hexane = 1:4
   Pour the developing solvent to the TLC chamber about 0.3 cm from the bottom.
3. Use a pencil to lightly draw a line across the plate about 0.6 cm from the bottom. Then mark three spots on this line which will indicate the position of the samples.
   The first mark is spotted with the first sample and the third with the second sample. The second mark is spotted with the first and the second sample together (co–spot). Put the resulting TLC plate in the TLC chamber standing on the end using forceps.
4. Develop the TLC plate. Before the solvent front reaches 0.5 cm from the top, remove the plate from the chamber and allow it to dry. [Note 2]
5. Illuminate the resulting plate with the ultraviolet light. Encircle each spot lightly with a pencil.
6. With the use of a ruler, measure the distances that the developing solvent and each compound have traveled. Using this data, calculate the R_f value of each compound.
7. Write down the conditions of the developing solvent and draw the resulting TLC plate. [Note 3]

Date of experiment : **Room temperature :**

Observation :

Name	Structural formula	M. W.	m.p. (℃)	b.p. (℃)	R_f* (E.A.:Hex.= :)
Benzyl alcohol					
Benzyl bromide					

R_f* : Write down the condition of the developing solvent when tracing the TLC plate.

〈Tracing the TLC plate〉

Checked by ____________________

Note

1. The R_f values are not only affected by the polarity but also by the ability of the compounds for hydrogen bonding. Benzyl alcohol has an oxygen–hydrogen bond which is available for hydrogen bonding on the silica gel.

2. Although the empirical formula for silica gel is SiO_2, its structure is based on the network of tetrahedral SiO_4 containing a single bond between an oxygen atom and a silicon atom rather than discrete SiO_2 molecules.

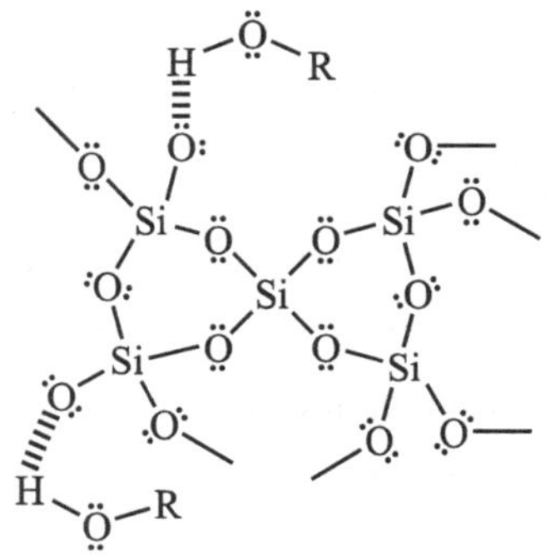

3. The anion, the cation and the radical at the benzylic position are relatively stable. The developed charge is spread over in the molecule instead of being concentrated on one carbon because of the delocalization due to the following resonance contributions.

* = cation, anion, radical

A similar behavior can be observed at the allylic position.

* = cation, anion, radical

A primary allylic or benzylic carbocation is as stable as a secondary alkyl carbocation. And a secondary allylic or benzylic carbocation is as stable as a tertiary alkyl carbocation.

Experiment 2-2

Experimental procedure

1. Preparation of the sample — put 50.0 mg of each sample in separate vials, and add 3.0 mL of ethyl acetate.

   a) aniline b) nitrobenzene c) acetanilide

2. Preparation of the developing solvent — Make the developing solvents in various ratios of ethyl acetate and hexane.

   Make developing solvents a) ethyl acetate : hexane = 1:4
   b) ethyl acetate : hexane = 1:2

   Pour the developing solvent to the TLC chamber about 0.3 cm from the bottom.

3. Use a pencil to lightly draw a line across the plate about 0.6 cm from the bottom. Then mark three spots on this line which will indicate the position of the samples.
   The first mark is spotted with the first sample and the third with the second sample. The second mark is spotted with the first and the second sample together (**co–spot**). Put the resulting TLC plate in the TLC chamber standing on the end using forceps.

4. Develop the TLC plate. Before the solvent front reaches 0.5 cm from the top, remove the plate from the chamber and allow it to dry.

5. Illuminate the resulting plate with the ultraviolet light. Encircle each spot lightly with a pencil.

6. Measure the distances that the developing solvent and each compound have traveled with a ruler. Using this data, calculate the R_f value of each compound.

7. Write down the conditions of the developing solvent and draw the resulting TLC plate. [Note 3 or 4]

Date of experiment : **Room temperature :**

Observation :

Name	Structural formula	M. W.	m.p. (℃)	b.p. (℃)	R_f* (E.A:Hex.= :)	
					1 : 4	1 : 2
Aniline						
Nitrobenzene						
Acetanilide						

R_f* : Write down the condition of the developing solvent when tracing the TLC plate.

〈Tracing the TLC plate〉

Checked by ____________________

Note

1. Even though the polarity of the nitro group is higher than that of the amino group, the R_f value of nitrobenzene is greater than that of aniline. As discussed in the previous section, aniline has two nitrogen–hydrogen bonds that are available for hydrogen bonding on the silica gel and therefore it holds a low R_f value.

2. The developing solvent system should be more polar for acetanilide since its R_f value in the developing solvent system (ethyl acetate : hexane = 1:4) is too low.
   Try the developing system (ethyl acetate : hexane = 1:2).

3. Acetanilide is very polar because of the large resonance contribution by the overlapping between the lone–pair electrons on the nitrogen and the π bond in a carbonyl group.

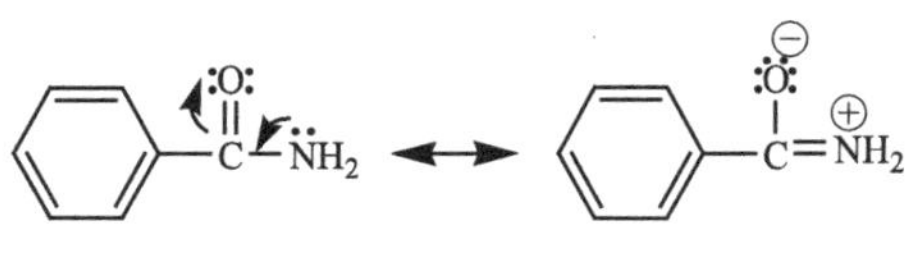

large contribution

In comparison, the same lone–pair electrons on oxygen in ester are not easily conjugated with the π bond in the carbonyl group because of the strong electronegativity of the oxygen atom.

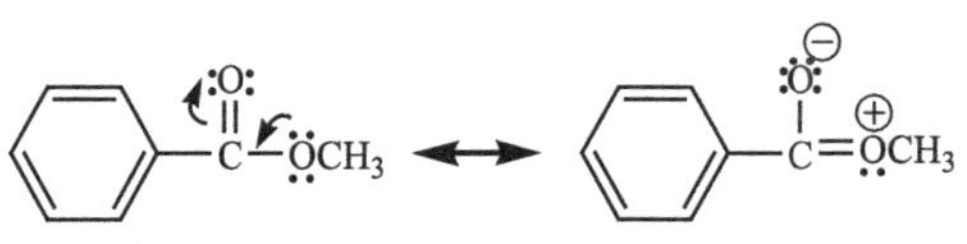

small contribution

4. Acetanilide also has a nitrogen–hydrogen bond, which is available for hydrogen bonding on the silica gel. For more explanation, Refer to the previous section **Note 2** on p37

Experimental procedure

Experiment 2-3

1. Preparation of the sample — put 50.0 mg of each sample in separate vials, and add 3.0 mL of ethyl acetate.
   a) benzoic acid b) methyl benzoate c) benzonitrile

2. Preparation of the developing solvent — Make the developing solvent in a ratio of ethyl acetate : hexane = 1:4. In order to prevent the tailing of benzoic acid, add 3~4 drops of formic acid to the developing solvent. Pour the developing solvent to about 0.3 cm of the TLC chamber from the bottom.

3. Use a pencil to lightly draw a line across the plate about 0.6 cm from the bottom. Then mark three spots on this line which will indicate the position of the samples.
   The first mark is spotted with the first sample and the third with the second sample. The second mark is spotted with the first and the second sample together (**co–spot**). Put the resulting TLC plate in the TLC chamber standing on the end using forceps.

4. Develop the TLC plate. Before the solvent front reaches 0.5 cm from the top, remove the plate from the chamber and allow it to dry.

5. Illuminate the resulting plate with the ultraviolet light. Encircle each spot lightly with a pencil.

6. Measure the distances that the developing solvent and each compound have traveled with a ruler. Using this data, calculate the R_f value of each compound.

7. Write down the conditions of the developing solvent and draw the resulting TLC plate. [Note 3]

Date of experiment : **Room temperature :**

Observation :

Name	Structural formula	M. W.	m.p. (℃)	b.p. (℃)	R_f* (E.A.:Hex.= :)
Benzoic acid					
Methyl benzoate					
Benzonitrile					

R_f* : Write down the condition of the developing solvent when tracing the TLC plate.

〈Tracing the TLC plate〉

Checked by ____________________

Note

1. The R_f value of benzoic acid is low since it has an oxygen–hydrogen bond and a polar carbonyl group. Meanwhile the R_f value of methyl benzoate is greater than that of benzoic acid since it does not have an oxygen–hydrogen bond.

2. When the acidic compounds are checked on TLC, a tailing or long strip is often observed on the TLC plate. It is caused by various salt formations since the acidic compound is sometimes deprotonated on the surface of the silica gel. Adding a small amount of a strong acid —formic or acetic acid—might reduce the tailing by preventing the acidic compounds from deprotonation.

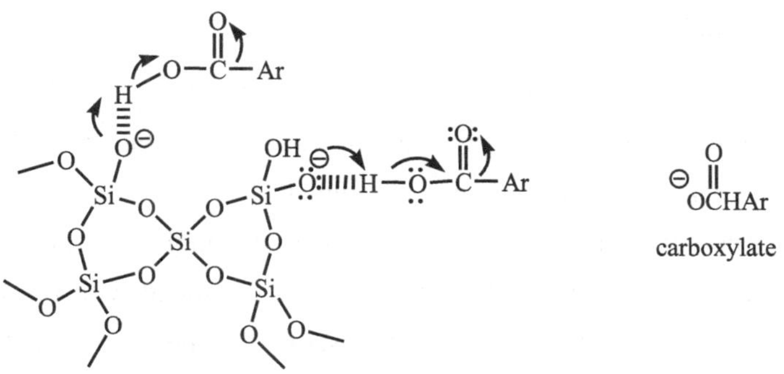

carboxylate

3. Nitrile and ester are derivatives of the carboxylic acid and have similar physical and chemical properties. For example, the melting point of benzonitrile is –13℃ and that of methyl benzoate is –12℃. The boiling point of benzonitrile is 191℃ and that of methyl benzoate is 198~199℃.

Experimental procedure

1. Preparation of the sample — put 50.0 mg of each sample in separate vials, and add 3.0 mL of ethyl acetate.

   a) diethyl maleate / diethyl fumarate
   <the developing solvent is ethyl acetate : hexane = 1 : 4>

   b) *cis*–stilbene / *trans*–stilbene
   <the developing solvent is ethyl acetate : hexane = 1 : 8>

2. Preparation of the developing solvent — Make developing solvents in various ratios of ethyl acetate and hexane. (1:4 and 1:8). Pour the developing solvent to about 0.3cm of the TLC chamber from the bottom.

3. Use a pencil to lightly draw a line across the plate about 0.6 cm from the bottom. Then mark three spots on this line which will indicate the position of the samples.
   The first mark is spotted with the first sample and the third with the second sample. The second mark is spotted with the first and the second sample together (**co–spot**). Put the resulting TLC plate in the TLC chamber standing on the end using forceps.

4. Develop the TLC plate. Before the solvent front reaches 0.5 cm from the top, remove the plate from the chamber and allow it to dry.

5. Illuminate the resulting plate with the ultraviolet light. Encircle each spot lightly with a pencil.

6. Measure the distances that the developing solvent and each compound have traveled. with a ruler. Using this data, calculate the R_f value of each compound.

7. Write down the conditions of the developing solvent and draw the resulting TLC plate. [Note 1,2]

Date of experiment : **Room temperature :**

Observation :

Name	Structural formula	M. W.	m.p. (℃)	b.p. (℃)	R_f* (E.A.:Hex.= :)
Diethyl maleate					
Diethyl fumarate					

R_f* : Write down the condition of the developing solvent when tracing the TLC plate.

⟨Tracing the TLC plate⟩

Name	Structural formula	M. W.	m.p. (℃)	b.p. (℃)	R_f* (E.A.:Hex.= :)
cis-Stilbene					
trans-Stilbene					

R_f* : Write down the condition of the developing solvent when tracing the TLC plate.

⟨Tracing the TLC plate⟩

Note

1. The dipole moments of the polar groups in the *trans* isomers cancel out each other since they are located on the opposite sides of the double bond. In the case of a *cis* isomer, the dipole moments caused by polar groups are additive since they are located on the same side of the double bond. In principle, *trans* isomers are less polar than *cis* isomers. As a result, the R_f value of diethyl fumarate is greater than that of diethyl maleate. The boiling point of diethyl fumarate (b.p. 218~219℃) is lower than that of diethyl maleate (b.p. 225℃).

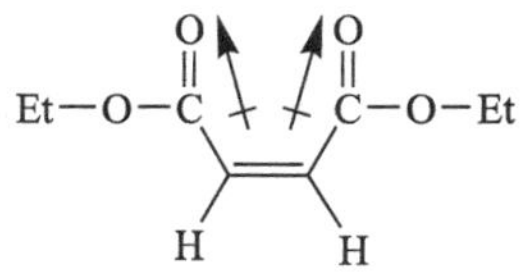

b.p. : 225℃

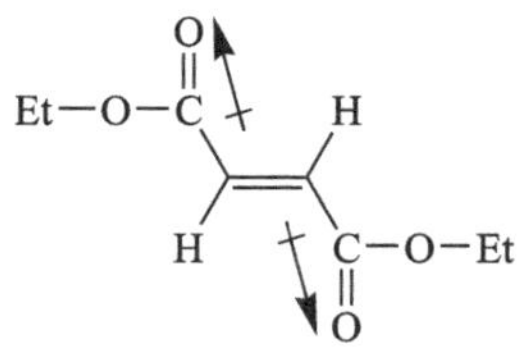

b.p. : 218~219℃

2. *cis*–Stilbene and *trans*–stilbene are exceptional. The R_f value of *cis*–stilbene is greater than that of *trans*–stilbene. It means that *cis*–stilbene is less polar than *trans*–stilbene. Two phenyl rings in *cis*–stilbene are not in the same plane because of the steric hindrance. The π electrons in phenyl ring of *cis*–stilbene are localized in its own ring. As a result, the instantaneous dipole moment in *trans*–stilbene is larger than that in *cis*–stilbene.

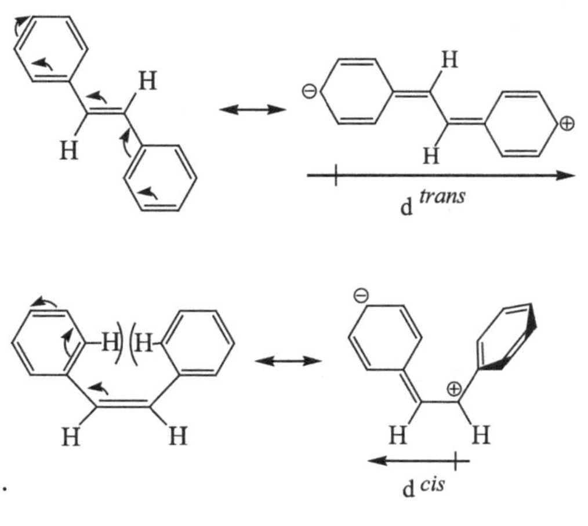

$d^{trans} > d^{cis}$

Experimental procedure

Experiment 3

1. Preparation of the sample — put 50.0 mg of each sample in separate vials, and add 3.0mL of ethyl acetate.
   a) *o*–dinitrobenzene / *m*–dinitrobenzene / *p*–dinitrobenzene
      <the developing solvent is ethyl acetate : hexane = 1 : 2>
   b) *o*–nitrobenzoic acid / *p*–nitrobenzoic acid
      <the developing solvent is ethyl acetate : hexane = 1 : 2>

2. Preparation of the developing solvent — Make a developing solvent in a ratio of ethyl acetate : hexane = 1:2. Pour the developing solvent to about 0.3 cm of the TLC chamber from the bottom.

3. Use a pencil to lightly draw a line across the plate about 0.6 cm from the bottom. Then mark three spots on this line which will indicate the position of the samples.
   The first mark is spotted with the first sample and the third with the second sample. The second mark is spotted with the first and the second sample together (**co–spot**). Put the resulting TLC plate in the TLC chamber standing on the end using forceps.

4. Develop the TLC plate. Before the solvent front reaches 0.5 cm from the top, remove the plate from the chamber and allow it to dry.

5. Illuminate the resulting plate with the ultraviolet light. Encircle each spot lightly with a pencil.

6. Measure the distances that the developing solvent and each compound has traveled with a ruler. Using this data, calculate the R_f value of each compound.

7. Write down the conditions of the developing solvent and draw the resulting TLC plate. [Note 1,2]

Date of experiment : **Room temperature :**

Observation :

Name	Structural formula	M. W.	m.p. (℃)	b.p. (℃)	R_f* (E.A.:Hex.= :)
o-Dinitrobenzene					
m-Dinitrobenzene					
p-Dinitrobenzene					

R_f* : Write down the condition of the developing solvent when tracing the TLC plate.

〈Tracing the TLC plate〉

Name	Structural formula	M. W.	m.p. (℃)	b.p. (℃)	R_f* (E.A.:Hex.= :)
o-Nitrobenzoic acid					
p-Nitrobenzoic acid					

R_f* : Write down the condition of the developing solvent when tracing the TLC plate.

〈Tracing the TLC plate〉

Note

1. *p*–Dinitrobenzene is less polar than *o*–dinitrobenzene based on the previous discussion about *cis* and *trans* isomerism. The dipole moments of the two nitro groups in *p*–nitrobenzene cancel out each other since they are located in the opposite direction – 180° – of the aromatic ring. As a result, the R_f value of *p*–nitrobenzene is greater than that of *o*–nitrobenzene. In contrast, *o*–nitrobenzene and *m*–nitrobenzene are more polar than *p*–nitrobenzene since the angles of the nitro groups are 60° and 120°, respectively.
Overall, the order of polarity is *ortho* > *meta* > *para* and the order of R_f value is *ortho* < *meta* < *para.*

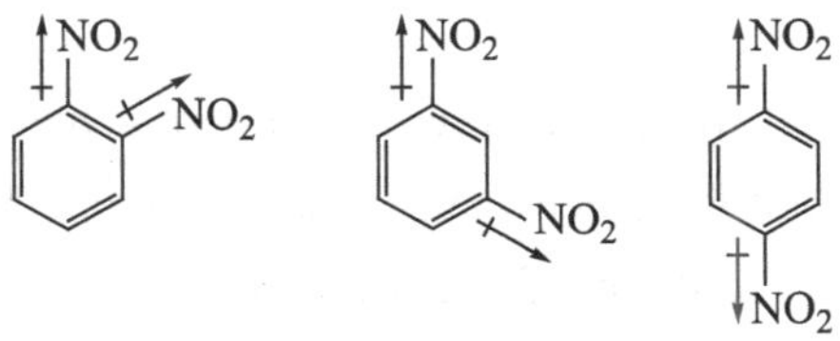

2. The same principle can be applied to *o*–nitrobenzoic acid and *p*–nitrobenzoic acid since both the carboxyl group and the nitro group of nitrobenzoic acid are strong electron withdrawing groups (EWG).

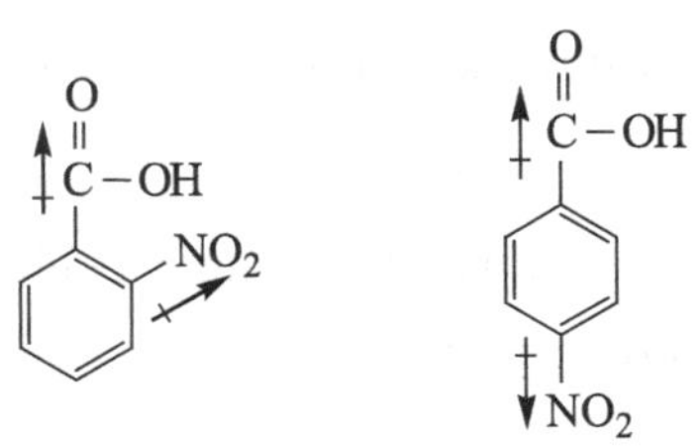

Experiment 3

Experimental procedure

1. Preparation of the sample — put 50.0 mg of each sample in separate vials, and add 3.0mL of ethyl acetate.
   a) *o*–nitrophenol / *p*–nitrophenol
      <the developing solvent is ethyl acetate : hexane = 1 : 2>
   b) *o*–nitroaniline / *p*–nitroaniline
      <the developing solvent is ethyl acetate : hexane = 1 : 2>

2. Preparation of the developing solvent — Make the developing solvent in a ratio of ethyl acetate : hexane = 1:2. Pour the developing solvent to about 0.3 cm of the TLC chamber from the bottom.

3. Use a pencil to lightly draw a line across the plate about 0.6 cm from the bottom. Then mark three spots on this line which will indicate the position of the samples.
   The first mark is spotted with the first sample and the third with the second sample. The second mark is spotted with the first and the second sample together (**co–spot**). Put the resulting TLC plate in the TLC chamber standing on the end using forceps.

4. Develop the TLC plate. Before the solvent front reaches 0.5 cm from the top, remove the plate from the chamber and allow it to dry.

5. Illuminate the resulting plate with the ultraviolet light. Encircle each spot lightly with a pencil.

6. Measure the distances that the developing solvent and each compound has traveled with a ruler. Using this data, calculate the R_f value of each compound.

7. Write down the conditions of the developing solvent and draw the resulting TLC plate. [Note 1,2]

Date of experiment : **Room temperature :**

Observation :

Name	Structural formula	M. W.	m.p. (℃)	b.p. (℃)	R_f* (E.A.:Hex.= :)
o-Nitrophenol					
p-Nitrophenol					

R_f* : Write down the condition of the developing solvent when tracing the TLC plate.

〈Tracing the TLC plate〉

Name	Structural formula	M. W.	m.p. (℃)	b.p. (℃)	R_f* (E.A.:Hex.= :)
o-Nitroaniline					
p-Nitroaniline					

R_f* : Write down the condition of the developing solvent when tracing the TLC plate.

〈Tracing the TLC plate〉

1. The polarity of organic compounds can be estimated by the vector sum of the dipole moments of the electron donating groups (EDG) and the electron withdrawing groups (EWG). The typical EDG and EWG are as the following:

   EDG: amino (–NH_2), hydroxyl (–OH), methoxy (–OCH_3)

   EWG: nitro (–NO_2), nitrile (–CN), carboxyl (–COOH)

2. In general, the *para* isomers are less polar than the corresponding *ortho* isomers. They travel farther on TLC resulting in greater R_f values than those of *ortho* isomers.

   However, the isomers of nitrophenols show the opposite result on TLC. They are interesting molecules which contain both the strong electron donating amine group and the strong electron withdrawing nitro group together. The *p*–nitrophenol has a bigger dipole moment than the *o*–nitrophenol because of the resonance contribution that creates a large distance between the two charges. Therefore, the R_f value of *o*–nitrophenol is greater than that of *p*–nitrophenol.

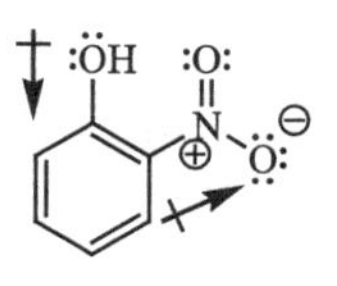

$\varepsilon = 16.30$

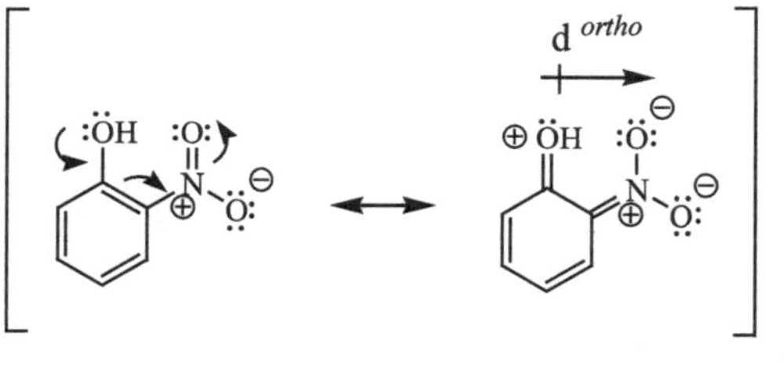

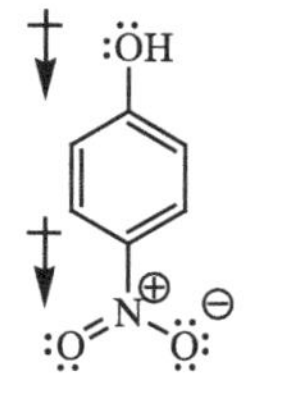

$\varepsilon = 42.20$

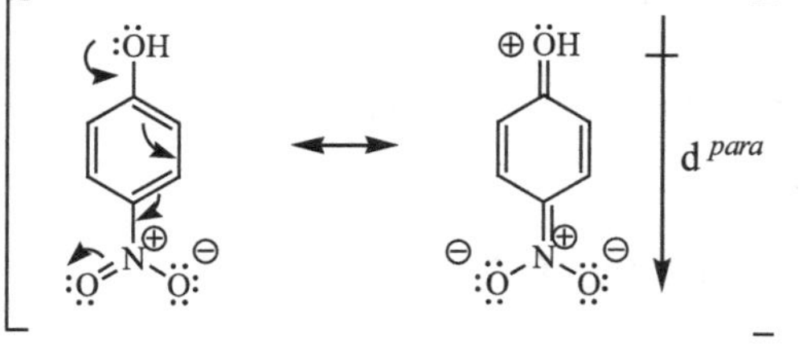

d *ortho* < d *para*

3. *p*–Nitroaniline also has both the strong electron donating group (–NH_2) and the strong electron withdrawing group (–NO_2). Based on the same discussion for nitrophenols, *o*–nitroaniline holds a greater R_f value than *p*–nitroaniline on TLC.

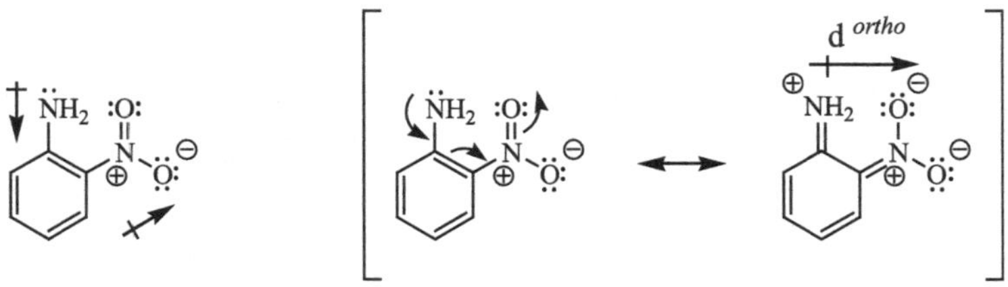

$\varepsilon = 47.30$

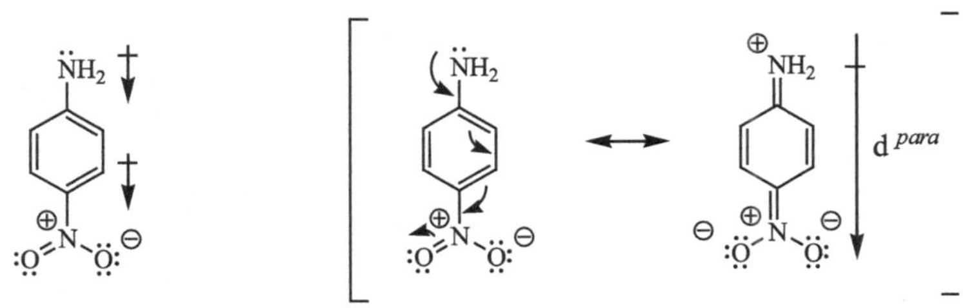

$\varepsilon = 78.30$

$$d^{ortho} < d^{para}$$

Keywords for Thin Layer Chromatography

Summary for Isomers

1.3 Distillation and Polarizability

Keywords : distillation, evaporation, reflux, boiling point, van der Waals, polarizability, azeotrope

Distillation has been used to increase the alcoholic content of fermented fruits or crops. Fractionation of gasoline (b.p. : 30–200℃), kerosene (b.p. : 175–300℃) and heating oil (b.p. : 275–400℃) from the crude oil using the difference in their boiling points is another good example of distillation.

Distillation is a basic technique for chemical experiments, which depends on the difference in the vapor pressure of the components of a liquid mixture. Liquid compounds can be purified and isolated by distillation since the vapor above the liquid mixture is enriched with a more volatile component. In distillation, liquid is heated to its **boiling point** and vaporized in the distillation apparatus. Then the hot vapor is condensed by cooling and the resulting liquid is collected in a receiving flask. When a volatile liquid is distilled off and the residue is obtained for a further process, it is called **evaporation**. Especially, the evaporation under reduced pressure through a rotary evaporator is widely used to remove the common organic solvent in the organic laboratory.

distillation

reflux

Reflux is based on the fact that the temperature of a liquid cannot exceed its boiling point when it is boiled in a flask equipped with the water–cooled condenser. In a chemical reaction, reflux is applied to maintain a certain temperature (the boiling point of a solvent).

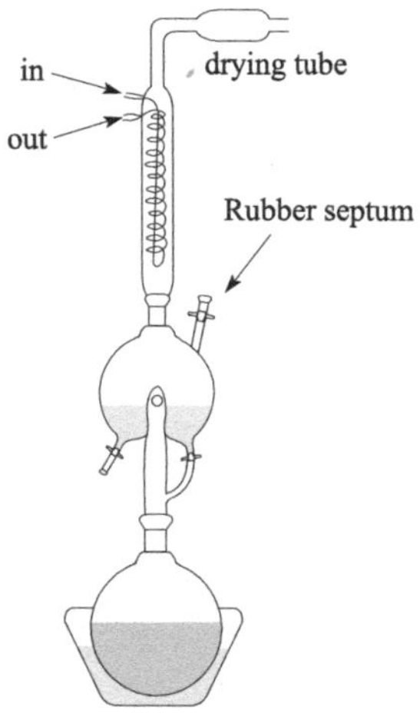

reflux and distillation set for anhydrous solvent

It is important to notice that the level between the solvent in the flask and the heat source is different for distillation and reflux. In the case of reflux, the level of the solvent should be higher than the heat source to prevent the reaction mixture from overheating by the resulting vapor.

The combination of reflux and distillation is applied to remove a trace of water from organic solvents. Organic solvents are refluxed with Na metal until water is completely removed and then distilled for the anhydrous reaction.

The **boiling point** of a liquid is defined as the temperature at which its vapor pressure equals the atmospheric pressure. In the laboratory, the observed boiling temperature actually occurs a few degrees above or below the literature value because of the different experimental conditions (For example, the weather and the atmospheric pressure are varied according to the location.)

Even when the compound being distilled is pure, a small portion of liquid will always be left in the flask at the end of the distillation as a result of condensation of the vapor in the flask upon cooling. **The flask containing substances to be distilled should never be heated to dryness** since there is a possibility that the glass flask will be cracked. Heating should be stopped just before the last traces of the liquid have been vaporized in order to avoid the decomposition and charring in the flask.

In general, a compound with a high boiling point holds greater attraction force between the molecules and *vice versa*. Therefore, the boiling point provides valuable experimental information about the intermolecular attraction force, which is closely related to the physical and chemical properties of the organic molecules. Here we deal with **van der Waals interaction** for the nonpolar compound and the **dipole–dipole interaction** for the polar compound, respectively.

I. Nonpolar molecules and van der Waals interaction

The boiling point of a nonpolar compound is usually low since the intermolecular interaction between the nonpolar molecules is small. However, as the size of the molecule gets bigger, its intermolecular force increases rapidly to attract each other. For example, hydrocarbons containing 1 to 4 carbons are gas, 5 to 17 are liquid and over 18 are solid under 1 atm pressure.

The weak attraction forces that exist between the nonpolar molecules are called **van der Waals interactions**. The nonpolar molecule that does not contain the permanent dipole can produce an **instantaneous dipole moment** by the electron movement within bonds or by external influences.

The instantaneous dipole polarizes the electron cloud of the nonpolar molecule into positive and negative electric charges and makes up the **induced dipole**. The degree of distortion in the electron cloud caused by fluctuating the induced dipoles is called **polarizability**.

When a partial positive electric charge of a molecule is aligned with another molecule's partial negative electric charge, the whole energy in the system decreases as both positive and negative electric charges tend to attract each other. The attraction between these induced dipoles is called **London dispersion force**. Although the induced dipoles are small and constantly changing, the net result provides a considerable amount of attraction force between the molecules. As the number of carbon and hydrogen atoms increases, the additive effect of these weak intermolecular forces becomes more significant, as evidenced by the increase in the boiling point from methane (–162℃) to hexane (69℃).

Polarizability which mainly depends on the size and shape of a molecule is very important in understanding both physical and chemical properties of organic compounds–especially nonpolar compounds.

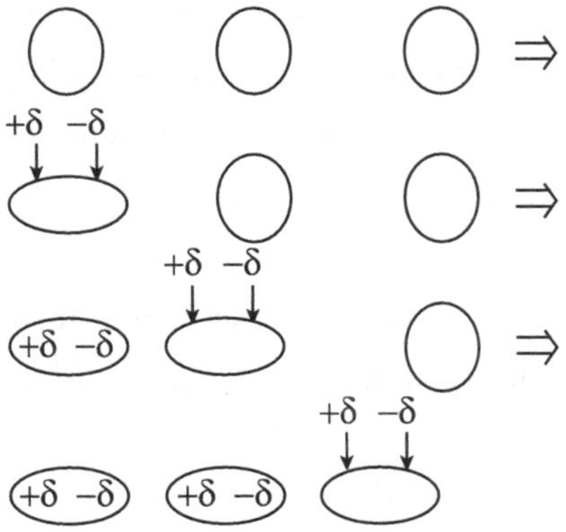

a. Nonpolar molecules which do not have a dipole moment.

b. They can create an instantaneous dipole by outside interference or electron movement in itself.

c. Instantaneous dipole makes the induced dipole by the polarization of other nonpolar molecules.

II. Correlation between the boiling point and the size and shape of the molecule

1) Bigger molecules which hold bigger electron clouds have loosely bound outer electrons compared to the smaller ones. They are susceptible to form polarized electric charges and create large induced dipoles compared to the tightly bonded electrons in smaller molecules. Therefore, bigger molecules have greater polarization than the smaller ones. For example, methane boils at –162℃, but

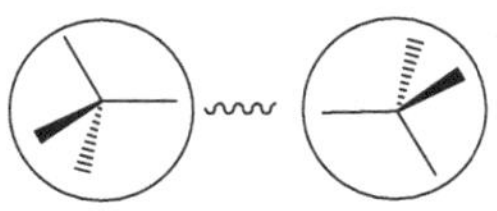

carbon tetrachloride which is substituted by a bigger atom–chlorine instead of hydrogen–boils at 77℃ and carbon tetrabromide boils at 190℃.

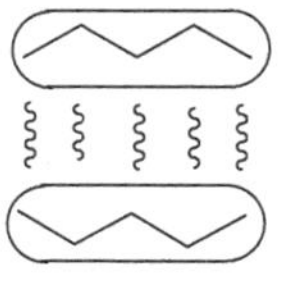

2) In general, the linear shape molecules have higher boiling points than the spherical shape molecules since they have more surfaces to interact with each other. For example, n–pentane boils at 35℃, while neopentane boils at 9.5℃.

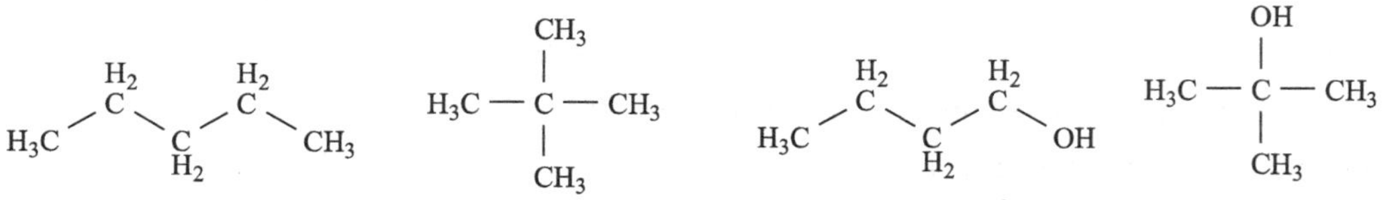

n-Pantane (35°C) Neopentane (9.5°C) n-Butanol (116-118°C) t-Butanol (85°C)

III. Polar molecules and dipole–dipole interaction

Since the dipole–dipole interaction has been dealt thoroughly in **1.1 Dissolution**, its effects on the boiling points will be dealt briefly in this chapter. In general, the more polar the compound, the higher its boiling point. Conversely, the compound which has a higher boiling point is more polar than the one with a lower boiling point.

1) The boiling point of water and ammonia (NH_3) is much higher than that of methane (CH_4) because of their polarity and ability to form the hydrogen bonding.

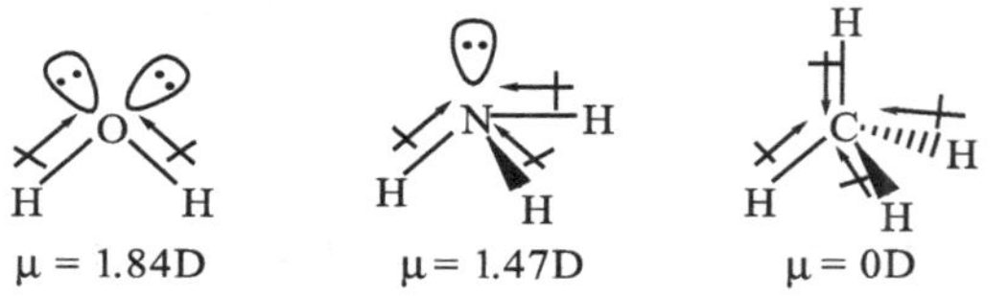

μ = 1.84D 100℃ μ = 1.47D -33℃ μ = 0D -162℃

2) The *cis* isomer (*Z*–isomer) is usually more polar than the *trans* isomer (*E*–isomer). For example, the boiling point of dimethyl maleate is 225℃ which is about 7℃ higher than that of dimethyl fumarate.

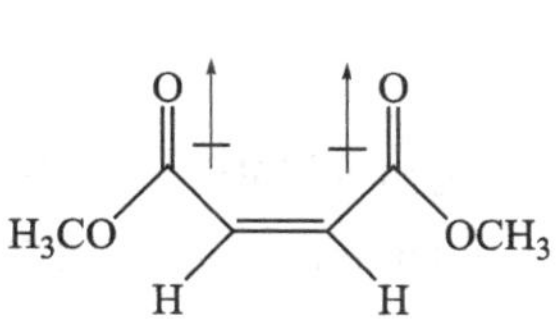

Dimethyl maleate (225℃)

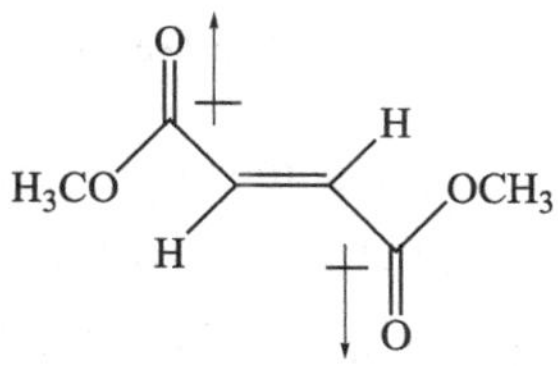

Dimethyl fumarate(218~219℃)

3) 1,3–pentadiene(piperylene) has a boiling point over 16℃ higher than unconjugated 1,4–pentadiene. When double bonds are conjugated, the contribution of resonance contributor I increases the molecular dipole since the distance between the opposite charges increases.

1, 4-Pentadiene (26℃)

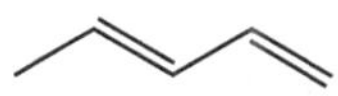

1, 3-Pentadiene (42℃)
(piperylene)

(I)

d^{uc} < d^{c}

4) Cyclic compounds are more polar than the corresponding acyclic compounds. They hold higher boiling points. (refer to **1.1 Dissolution**)

Acyclic compound	cyclic compound
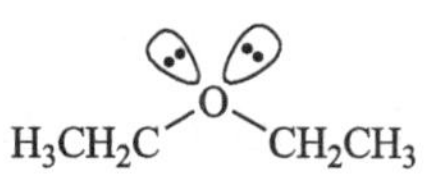 Diethyl ether (34.6℃)	 THF (65℃)
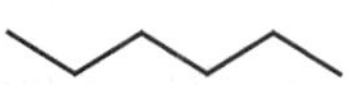 Hexane (68.7℃)	 Cyclohexane (80.7℃)

IV. Polarity v.s. Polarizability

When comparing the dipole moments of hydrochloric acid (HCl) and hydroiodic acid (HI), HI (0.448 D) is less polar than HCl (1.109 D). HI is expected to hold a lower boiling point considering its lower polarity. But the boiling point of HI (–35.1℃) is higher than that of HCl. (–85.0℃)

Compound	b.p.	pK_a	Difference in electronegativity	Dipole moment (D)	Anion radius (Å)
HCl	-85.0℃	-7	3.0-2.1 = 0.9	1.109	1.81
HI	-35.1℃	-9	2.4-2.1 = 0.3	0.448	2.16

In fact, the boiling point is not only affected by polarity but also by polarizability. The higher boiling point of HI can be explained by polarizability since HI molecule (2.16 Å) is bigger than HCl (1.81 Å) As a result, polarizability sometimes has priority over polarity in explaining some physical property of organic compounds. The verification of this explanation can be double checked by acidity of HI and HCl. In terms of the anion stability, an chlorine anion should be more stabilized by polarity owing its high electronegativity (3.0). However, a large iodine anion is stabilized by polarizability in addition to polarity (EN= 2.4). In fact, HI (–9) is a stronger acid than HCl (–7).

Polarizability and polarity are observed in both polar or nonpolar organic compounds. They complement each other to elucidate the **relationship between the physical property and the structure of organic compounds**. Now it is important to understand how and when they are applied in organic chemistry.

In this experiment, a simple distillation and removal of water using Dean–Stark water separator set are introduced.

Experiment I

Reagents

Name	Structural formula	M. W.	m.p. (℃)	b.p. (℃)	Density (g/mL)
Aniline					
H_2O					

Apparatus

⇨ magnetic stirrer, magnetic spin-bar, lab jack, heating mentle, transformer

Glassware

⇨ distillation set, graduated cylinder, 100 mL round bottom flask, thermometer

Setting the experiment equipment

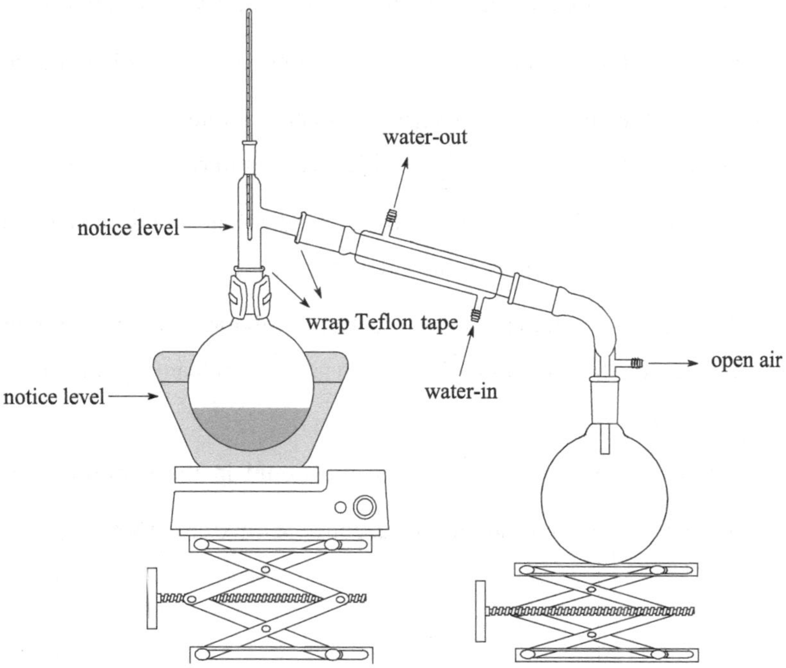

Experimental procedure

1. In a dry 100 mL round bottom flask, place 30.0 mL of aniline and 10.0 mL of water. [Note 1] Add a magnetic spin bar and install a distillation set. [Note 2] Do not over tighten the grounded glass joints. Wrap around each joint with Teflon tape. [Note 3] (Observation 1: state of mixed solvent)

   Make sure that the liquid level of the reaction mixture is the same or slightly lower than the level of the heat source.

2. **The distillation assembly must have an opening to atmosphere in order to release a high vapor pressure within the system when heat is applied.**

3. Start distillation with the transformer at 70V while continuously stirring.
   When distillation starts, check the temperature every 2 minutes. If a special change is observed, record the temperature and time. [Note 4] (Observation 2: change in temperature)

4. Record the time and the temperature when the first drop is placed on the thermometer and in the receiving flask, respectively.

   [Note 5] (Observation 3: color and state of the distilled liquid)

5. Check the temperature every 2 minutes. [Note 6]
   (Observation 4: change in distilled temperature)

6. Discontinue the distillation when the distillate condensed on the thermometer becomes transparent or clear.

7. Measure the volumes of the azeotropic mixture and the remaining aniline in the distilling flask by using a graduated cylinder.

   [Note 7]
   (Observation 5: color change of the liquid in the distilling flask)

8. Find the weight ratio of the azeotropic mixture. [Note 8,9]

Date of experiment: **Room temperature:**

Observation

Observation 1 :

Observation 2 :

Observation 3 :

Observation 4 :

Observation 5 :

- Theoretical yield ____________ g
- Practical yield ____________ g
- Yield ____________ %

Checked by ____________________

Note

1. When using aniline, make sure it is handled under the hood.
   The liquid mixture to be distilled should occupy around one–third to two–thirds of the distilling flask. If the bulb is filled more than two–thirds, there is danger that some of the liquid may boil over into the receiving flask. If the bulb is filled less than one–third, there will be an unnecessarily large heat loss resulting from the relatively large volume of vapor that is required to fill the distilling flask.

2. The spin-bar should be added to help minimize the formation of localized hot spots that may cause bumping.

3. When the distillation assembly has been completed, it should be checked for physical stability. Teflon tape is used to ensure the tightness of all joint connection. In the case of leaking, Teflon tape will be wetted by the distilling liquid and/or the vapor.

4. The top of the mercury bulb should be 5–10 mm below the opening of the side arm.
   The temperature will rise rapidly just before the distillation starts, so observe closely when the liquid boils in the distilling flask.

5. The proper way of supplying heat for distillation is to provide enough heat so that the liquid is distilled regularly at a uniform rate. Insufficient heating will temporarily stall the distillation and permit the bulb of the thermometer to cool below the distilling temperature, resulting in erratic temperature readings.

6. When a substance containing a small amount of impurities is distilled, the first portion of the distillate (the **forerun**) will contain more volatile impurities and a certain amount of the main liquid. As the temperature continues to rise, the bulk of the principal liquid will then distill over a short temperature range, usually 2–3℃ (the **principal** or **main fraction**). After this fraction has been distilled, the boiling point will rise or fall rapidly because of the presence of less volatile impurities (the **after–run**) or because no more vapors are

available, respectively.

7. The state and the color of the distilled liquid are different according to its composition at a given temperature. At the beginning, it is a cloudy or milky liquid that contains water and aniline.

8. The distilled mixture is divided into an aqueous solution and an organic layer since the mixed vapor is condensed to a liquid at room temperature. The upper organic layer might be cloudy because of a trace amount of water dissolved. The leftover aniline in the distilling flask turns light brown after the distillation.
   When the volume ratio of each layer is converted to the weight ratio, the composition of the distilled liquid at a given temperature can be obtained.

9. An **azeotrope** is a constant–boiling mixture with a definite composition. Because of the **azeotrope** formation, a 100% pure substance cannot be obtained by simple distillation from a miscible mixture. For example, a mixture of ethanol and water boils at 78.2℃ under 1 atm pressure. The composition of the distilled mixture is always H_2O : ethanol = 95.6 : 4.4. In principle, no alcohol that contains over 95.6% of ethanol can be obtained from the fermented fruit or crop by simple distillation.
   The boiling point of an azeotrope is always outside the range of the boiling point of each component. When the intermolecular force between the different components is smaller than that of the same molecule, the combined vapor pressure of the mixture at a certain composition is greater than that of the component. Therefore, the azeotropic mixture with a specific composition boils at a lower temperature –the **minimum boiling point**–. (Figure. 1) For example, when ethanol (78.3℃) and chloroform (61.2℃) are distilled together, the azeotropic mixture (59.4℃) – ethanol : chloroform = 7 : 93 – is obtained until one of the components is completely removed from the distillating flask.

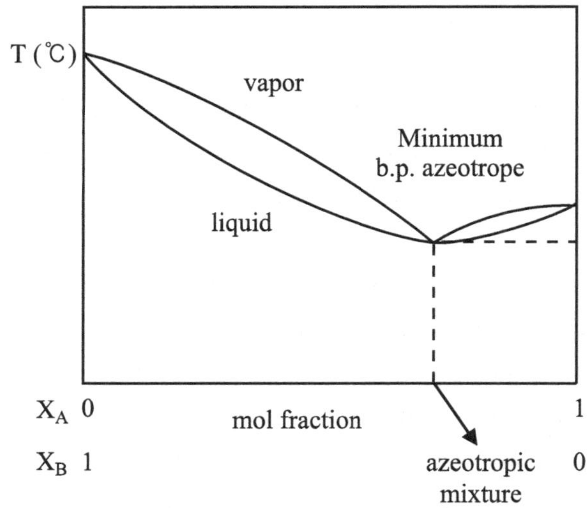

figure 1. Positive deviation of minimum–boiling point

When the intermolecular force between the different components is greater than that of the same molecule, the combined vapor pressure of the mixture at a certain composition is smaller than that of each component. The azeotropic mixture with a specific composition boils at a higher temperature –the **maximun boiling point** – than each component. (Figure. 2) In this case, the component that has a lower boiling point is distilled first and then the azeotropic mixture is distilled. For example, when formic acid (100.8℃) is mixed with water (100℃), the azeotrope mixture – formic acid : water = 77.4: 22.6 boils at 107.1℃.

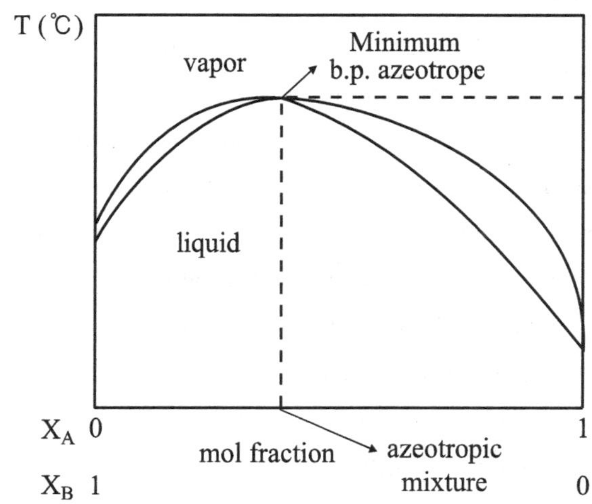

figure 2. Negative deviation of maximum–boiling point

Experiment 2

Reagents

Name	Structural formula	M. W.	m.p. (℃)	b.p. (℃)	Density (g/mL)
Toluene					
H_2O					

Apparatus

⇨ magnetic stirrer, magnetic spin-bar, lab jack, heating mentle, transformer

Glassware

⇨ Dean–Stark separatory, reflux condenser, graduated cylinder, 100mL round bottom flask, thermometer

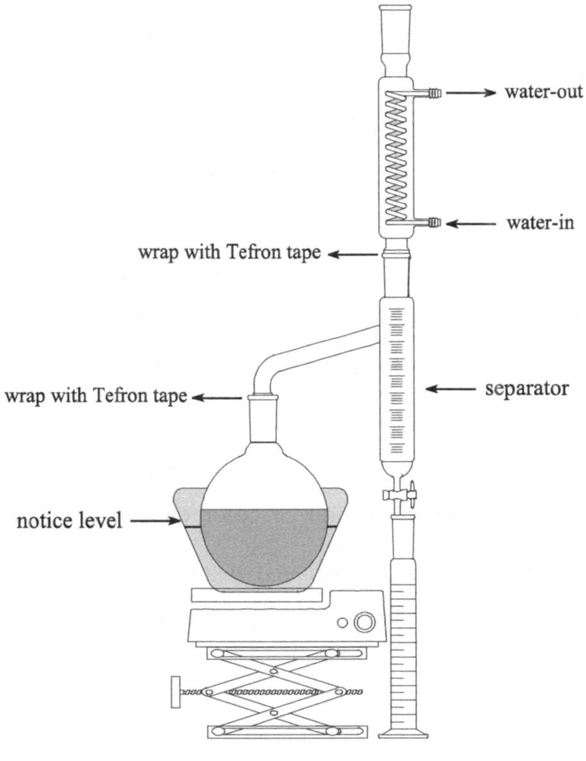

Dean–Stark water separator set

Experimental procedure

1. In a 100 mL round bottom flask, place 50.0 mL of toluene and 5.0 mL of water exactly. [Note 1]
   (Observation 1: actual volume of each sample)
2. Place the magnetic spin bar and install the Dean–Stark distillation set. Wrap around the connection joints with Teflon tape. [Note 2] **A reflux assembly must have an opening to the atmosphere to release a high vapor pressure within the system when heat is applied.**
3. Make sure that the liquid level of the reaction mixture is slightly higher than that of the heat source. Set the transformer to 70 V and start reflux while stirring.
4. When the liquid in the separator divides into two layers, collect the bottom layer in a graduated cylinder. [Note 3]
   (Observation 2: state of solvent in the condenser)
5. Continue reflux until the liquid condensed in the condenser of Dean–Stark distillation set turns transparent or clear. [Note 4]
   (Observation 3: state of solvent in the condenser)

7. Remove the heat source and cool the Dean–Stark distillation set. [Note 5]
8. Measure the volume of the liquids remaining in the distillating flask and collected from the separator. [Note 6]
   (Observation 4: volume of each solvent)
9. Add blue Drierite ($CaSO_4$ anhydride) to the solvents in the distilling flask and in the separator to check if any moisture is left. [Note 7] (Observation 5: color change on Drierite)
10. Find the yield of the isolated water. [Note 8,9,10]
    (Observation 6: yield of the aqueous solution)

Date of experiment : **Room temperature:**

Observation

Observation 1 :

Observation 2 :

Observation 3 :

Observation 4 :

Observation 5 :

Observation 6 :

- Theoretical yield ____________ g
- Practical yield ____________ g
- Yield ____________ %

Checked by ______________________

Note

1. This is an experiment to remove water from the mixture of toluene and water. In order to calculate the exact amount of water to be removed, the initial amount of toluene and water should be measured accurately.

2. When placing the liquid mixture in the Dean–Stark distillation flask, make sure the mixture in the round bottom flask does not exceed over 2/3. The spin-bar takes the role of a boiling chip that prevents bumping caused by the localized heat spot. Water is introduced from the lower end of the condenser and directed outward at the top of the condenser. Adjust the water pressure just enough to fill the condenser and flow constantly. Teflon tapes ensure the tightness of the connection joint. In the case of leaking, it will be wetted by the distilling liquid and /or the vapor.

3. The lower layer in the separator should be collected in the graduated cylinder before it overflows back into the distilling flask.

4. Because the vapor contains water and toluene that are immiscible, a cloudy or milky liquid will be condensed in the reflux condenser.

5. When the water is removed completely, only the pure toluene is refluxing. If it stays transparent or clear, continue the reflux for another 10 minutes to ensure the complete removal of water.

6. Lowering the temperature facilitates the separation of the resulting solution in the separator into two layers. The collected solvent is cooled at room temperature in order to maximize the separation between water and toluene. When the organic layer is still cloudy, add a small amount of salt to remove any traces of water. (**Salting–out effect)** Convert the measured volumes of water and remaining toluene into weights in order to calculate the ratio of the azeotrope.

7. The **Drierite** acts as an indicator for moisture. It turns from sky blue to light pink when it absorbs moisture. By observing the color change, the existence of moisture can easily be detected.

8. When 5.0 mL of water is used, about 4.5 mL can be separated.

9. An **azeotrope** is a constant–boiling mixture of a definite composition. The boiling point of an azeotrope is always outside the range of the boiling points of its components. For example, the minimum–boiling azeotropic mixture of toluene (b.p. : 110.6℃) and water boils at 84.1℃. For more explanation about azeotrope, Refer to the previous section **Note 9** on page 65.

10. Some common azeotropes

The azeotropic mixture in the mixture mixed with water	b.p. (℃)	Composition, wt % (water : others)
diethyl ether	34.6	1.3 : 98.7
ethanol	78.2	4 : 96
ethyl acetate	70.4	8.5 : 91.5
pentane	34.6	1.4 : 98.6
hexane	61.6	5.6 : 94.4
toluene	84.1	13.5 : 86.5
benzyl alcohol	99.9	91 : 9
aniline	98.6	80.8 : 19.2

Keywords for Distillation

Summary for Polarizability

1.4 Extraction and Acid Dissociation Constant (pK_a)

Keywords : distribution coefficient (K), salting-out, brine , pK_a

There are many examples of **extraction** in everyday life. Women remove manicure with acetone, an organic solvent. And as their makeup does not easily get off with water, they use cleansing products, which are also organic compounds. Grease-stained clothes are washed in the same way. These days perfumes are artificially synthesized. But in the past, grinded flowers were soaked in ether for a long time to separate the substance which is responsible for the odor and the color.

Extraction is one of the old techniques used in organic chemistry for the separation of valuable substances from materials that occur in nature, for the isolation of dissolved substances from solutions and also for the removal of soluble impurities. Extraction is a process of transferring a solute dissolved in a solvent to another solvent. The **solubility** of the solute for the second solvent should be much higher than that of the first one so that almost all of the solute could move to the second solvent. In addition, these two solvents do not mix together and produce two layers – the organic layer and the aqueous layer. The organic layer moves up or down according to the density of the solvent. Therefore, the density of the organic solvent must be checked before the experiment.

The concept of the **distribution coefficient** is based on the difference in solubility. The ratio of solubility of the solute in solvent 1 and 2 is called the distribution coefficient ($K = C_2/C_1$) where solubility is expressed in grams per milliliter. It is a constant value to each solute depending on the nature of the solvents at a given temperature.

$$K = \frac{C_2}{C_1} = \frac{\dfrac{\text{the amount of solute dissolved in solvent 2}}{\text{the volume of solvent 2}}}{\dfrac{\text{the amount of solute dissolved in solvent 1}}{\text{the volume of solvent 1}}}$$

The ratio of C_1 to C_2 is constant and independent to the total concentration and the amount used.

When the solubility of compound A is 10.0 g/100.0 mL H_2O (25℃)

and 100.0 g/100.0 mL ether, the value of K is 10.

$$K = \frac{\dfrac{100\ \text{g}}{100\ \text{mL of ether}}}{\dfrac{10.0\ \text{g}}{100\ \text{mL of water}}} = 10$$

When 5.0 g of compound A is dissolved in 100.0 mL of water, the results of (1) extracting it with 100.0 mL of ether and (2) extracting it twice with 50.0 mL of ether each time are as below.

Case (1)

First extraction

A = X when dissolved in 100mL of ether

A = 5.0 – X when dissolved in 100mL of water

$$K = 10 = \frac{C_1}{C_2} = \frac{\dfrac{X\ \text{g}}{100\text{mL of ether}}}{\dfrac{(5.0 - X)\ \text{g}}{100\text{mL of water}}}$$

$$K = 10 = \frac{X\ (100\text{mL})}{(5.0 - X)\ 100\text{mL}}$$

$5000 - 1000\ X = 100\ X$

$1100\ X = 5000$

$X = 4.55$

A = 0.45 g in water

4.55 g in ether

⇩

A = X

= 4.55 g (91.0 %)

when dissolved in the organic layer

Case (2)

First extraction

A = X when dissolved in 50mL of ether

A = 5.0 – X when dissolved in 100mL of water

$$K = 10 = \frac{C_1}{C_2} = \frac{\dfrac{X\ \text{g}}{50\text{mL of ether}}}{\dfrac{(5.0 - X)\ \text{g}}{100\text{mL of water}}}$$

A = 0.83 g in water

4.17 g in ether

Second extraction

A = Y when dissolved in 50mL of ether

A = 0.83 – Y when dissolved in 100mL of water

$$K = 10 = \frac{C_1}{C_2} = \frac{\dfrac{Y\ \text{g}}{50\text{mL of ether}}}{\dfrac{(0.83 - Y)\ \text{g}}{100\text{mL of water}}}$$

A = 0.14 g in water

0.69 g in ether

⇩

A = X + Y

= 4.17 g + 0.69 g

= 4.86 g (97.2 %)

when dissolved in the organic layer

As shown in the results above, it is better to use small portions of the second solvent several times than to make a single extraction with a large portion. This is the basis of washing or rinsing laboratory glassware at least three times to ensure their cleanness.

The **strength of acids** are expressed in **pK_a**, and its range is from –15 to 60. Standard inorganic acids such as sulfuric acid (H_2SO_4), nitric acid (HNO_3) and hydrochloric acid have pK_a values of –2 ~ –9, while typical organic acids have pK_a values of 3 ~ 20.

$$HA + H_2O \longrightarrow A^{\ominus} + H_3O^{\oplus}$$

$$K_{eq} = \frac{[H_3O^{\oplus}][A^{\ominus}]}{[HA][H_2O]} \qquad K_a = K_{eq}[H_2O] = \frac{[H_3O^{\oplus}][A^{\ominus}]}{[HA]}$$

$$pK_a = -\log K_a$$

The smaller the pK_a, the stronger the acid, and *vice versa*.

The pK_a value provides a criterion to determine the direction of the reaction as well as the strength of the acid.

$$\underset{pK_a = 5}{HA} + B^{\ominus} \rightleftharpoons A^{\ominus} + \underset{pK_a = 15}{HB}$$

As HA is a stronger acid than HB, it produces more stable anions, and therefore the forward reaction proceeds.

$$\underset{pK_a = 5}{R-\overset{\overset{O}{\|}}{C}-O\underline{H}} + {}^{\ominus}OH \rightleftharpoons R-\overset{\overset{O}{\|}}{C}-O^{\ominus} + \underset{pK_a = 16}{HO\underline{H}}$$

$$\underset{pK_a = 20}{R-\overset{\overset{O}{\|}}{C}-C\underline{H}_3} + (i\text{-}Pr)_2N^{\ominus}Li^{\oplus} \rightleftharpoons R-\overset{\overset{O}{\|}}{C}-\overset{\ominus}{C}H_2 + Li^{\oplus} + \underset{pK_a = 35}{(i\text{-}Pr)_2\ddot{N}\underline{H}}$$

In principle, the equilibrium moves to the direction of a higher pK_a value (from strong acid to weak acid) in all reactions.

The strength of acid HA depends on the stability of the anion(A^-). The more stable the anion, the stronger the acid.

	CH_4	NH_3	H_2O
pK_a	65	36	16

⟹ As the electronegativity increases, the anion becomes more stable and the pK_a value decreases.

$^{\ominus}CH_3$ $^{\ominus}NH_2$ $^{\ominus}OH$

	RCH_2CH_3	$RCH{=}CH_2$	$RC \equiv CH$
pK_a	65	45	25

⟹ As the S character increases, the anion becomes more stable and the pK_a value decreases.

$RCH_2-\overset{\ominus}{C}H_2$ (sp^3) $RCH{=}\overset{\ominus}{C}H$ $RC{\equiv}\overset{\ominus}{C}$

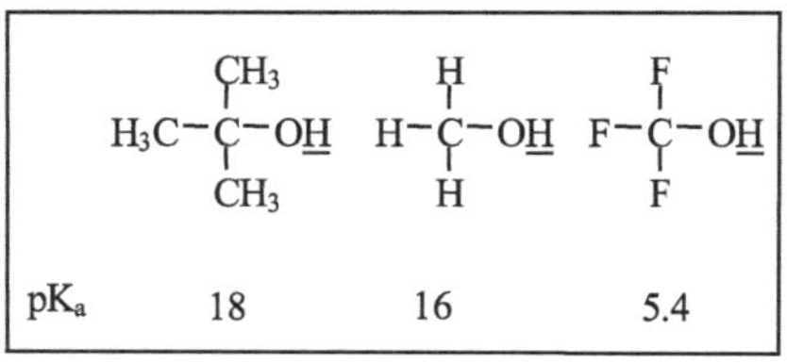

⟹ As the inductive effect increases, the anion becomes more stable and the pK_a value decreases.

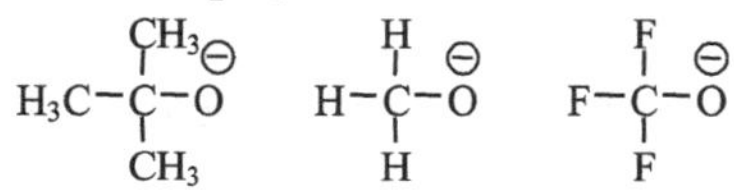

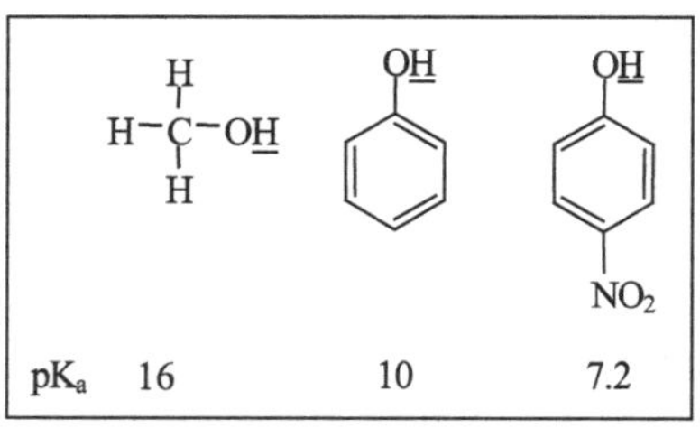

⟹ Resonance contribution and EWG stabilize the phenoxide anion to lower the pK_a.

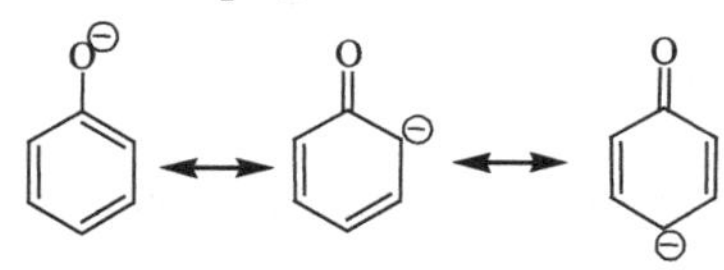

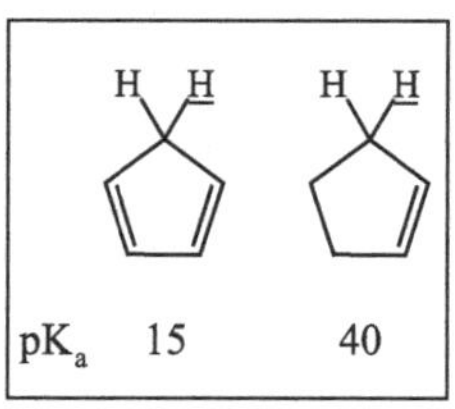

⟹ The anion of cyclopentadiene conjugates with the π-electron inside the ring and satisfies the Hückel rule (6 electrons). And the anion is stablized by the aromaticity to lower the pK_a.

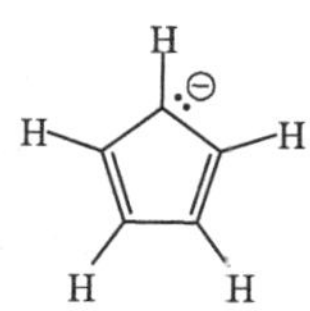

6 electrons

In this experiment, we separate each substance from a mixture by utilizing the difference in pK_a value as in theory. And the basic techniques for extraction is introduced.

Experiment

Reagents

Name	Molecular formula	Structural formula	M. W.	m.p. (℃)	b.p. (℃)	Density (g/mL)
Benzoic acid						
Benzyl alcohol						
Phenol						

Apparatus

⇨ magnetic stirrer, magnetic spin bar, lab jack, UV lamp

Glassware

⇨ 100 mL separatory funnel, 100 mL Erlenmeyer, TLC chamber, filter flask, Büchner funnel, glass rod, watch glass, pH paper

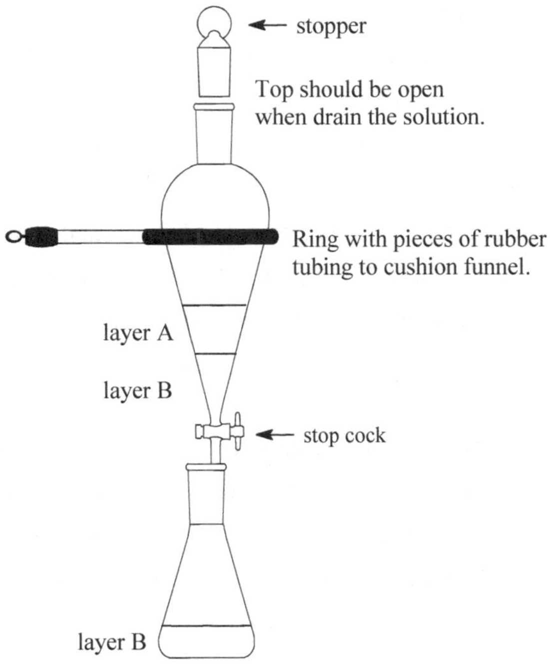

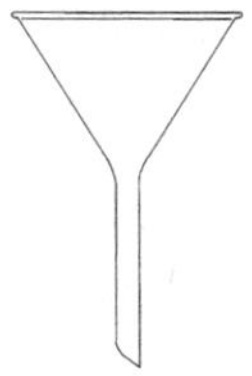

conical funnel

Experimental procedure [Note 1]

1. In a 100 mL Erlenmeyer flask **A,** place 1.50 g (12.3 mmol) of benzoic acid, 1.50 mL (1.57 g, 14.5 mmol) of benzyl alcohol and 1.00 mL (1.07 g, 11.4 mmol) of phenol in 10.0 mL of diethyl ether. [Note 2] To monitor the successive operations, prepare 3~4 TLC plates and spot them with the starting mixture. [Note 3]

<Separation of benzoic acid>

2. Add 20.0 mL of sat'd $NaHCO_3$ and stir for 2~3 minutes. [Note 4] (Observation 1: effervescence of CO_2)

3. Stop stirring and check the organic layer with the pre–spotted TLC. If benzoic acid is remained on the TLC, add 0.30 g of solid $NaHCO_3$ and stir for 5 minutes. Check TLC again. <The developing solvent is ethyl acetate : hexane = 1:3, R_f of benzoic acid = 0.27> [Note 5] (Observation 2: TLC)

4. After making sure benzoic acid has disappeared completely, transfer the solution from the 100 mL Erlenmeyer flask **A** to a 100 mL separatory funnel using a conical funnel. [Note 6]

5. Hold the separatory funnel firmly with both hands and shake it vigorously 2~3 times. Vent frequently with the lower end pointing upward and away from other people. Repeat shaking and venting until there is no "hissing" sound. When the funnel is placed in the ring, carefully remove the stopper. [Note 7] (Observation 3: gas evolution)

6. Allow the layers to stay undisturbed until the layers are separated completely. When the layers are separated, the lower layer is drawn off carefully through the stopcock into a 100 mL Erlenmeyer flask **B.** Rinse the stem with water. [Note 8] (Observation 4: separation of layers)

Date of experiment: **Room temperature:**

Observation

Observation 1 :

Observation 2 :

Observation 3 :

Observation 4 :

<Separation of phenol>

7. The upper layer is **poured through the mouth of the separatory funnel** into a clean Erlenmeyer flask **C**. [Note 9] Rinse the empty separatory funnel with 10.0 mL of diethyl ether.
   Add 20.0 mL (60.0 mmol) of 6.0 M sodium hydroxide solution (NaOH) and stir for 5 minutes. Make sure phenol has disappeared by checking the TLC of the organic layer. <the developing solvent is ethyl acetate : hexane = 1:3, R_f of phenol = 0.54> If phenol is remained, add 0.30 g of solid NaOH, stir for 5 minutes and check again with TLC. [Note 10] (Observation 5: separation of layers)

8. After making sure the phenol has completely disappeared, transfer the solution from the Erlenmeyer flask **C** to a 100 mL separatory funnel using a conical funnel. [Note 6]

9. Hold the separatory funnel firmly with both hands and shake it vigorously 2~3 times. Vent it frequently with the lower end pointing upward and away from other people. Repeat shaking and venting until there is no "hissing" sound. When the funnel is placed in the ring, carefully remove the stopper. [Note 7] (Observation 6: gas evolution)

10. Allow the layers to stand undisturbed until the layers are separated completely. When the layers are separated, the lower layer is drawn off carefully through the stopcock into a 100 mL Erlenmeyer flask **D.** Rinse the stem with water.
    (Observation 7: separation of layers.)

<Recovering benzyl alcohol>

11. Add 20.0 mL of brine to the organic layer in the separate funnel and follow the same procedure to separate the organic layer and the aqueous layer. Add 10.0 mL of brine and wash it once again. [Note 11] The lower layer is drawn off carefully through the stopcock into a 100 mL Erlenmeyer flask **E.**

Observation 5 :

Observation 6 :

Observation 7 :

12. The organic layer is **poured through the mouth of the separatory funnel** into a clean Erlenmeyer flask **F**. Rinse the empty separatory funnel with 10.0 mL of diethyl ether.
    Add a small amount of anhydrous magnesium sulfate ($MgSO_4$) to the organic layer in an Erlenmeyer flask **F** until the solution becomes clear and the solid $MgSO_4$ moves freely in the solution. [Note 12] (Observation 8: state of organic layer)

13. After water is completely removed, filter magnesium sulfate with suction on the Büchner funnel. Then rinse it twice with 10 mL of diethyl ether. [Note 13]

14. Transfer the resulting solution to a pre–weighed 100 mL round–bottom flask. Evaporate diethyl ether by using a rotary evaporator. When the solvent is almost evaporated, benzyl alcohol will remain at the bottom of the flask. [Note 14]
    Weigh the flask and check the purity of the product by TLC. Calculate the recovered amount of benzyl alcohol and the yield. (Observation 9: weight of benzyl alcohol)

<Recovering phenol>

15. Keep adding 6.0 M hydrochloric acid to the phenol in Flask **D** in small portions at a time until the pH of the solution becomes 1 on the pH paper. [Note 15] Transfer the solution to the separatory funnel using a conical funnel, add diethyl ether (extract twice with 10.0 mL), separate the layers and obtain the organic layer.

16. Add 20.0 mL of brine to the organic layer in the separate funnel and follow the same procedure to separate the organic layer and the aqueous layer. Add 10.0 mL of brine and wash it once again. [Note 11] The lower layer is drawn off carefully through the stopcock into a 100 mL Erlenmeyer flask **G**.

Observation 8 :

Observation 9 :

17. The organic layer is **poured through the mouth of the separatory funnel** into a clean Erlenmeyer flask **H**. Rinse the empty separatory funnel with 10.0 mL of diethyl ether
Add a small amount of anhydrous magnesium sulfate ($MgSO_4$) to the organic layer in the Erlenmeyer flask **H** until the solution becomes clear and the solid $MgSO_4$ moves freely in the solution [Note 12] (Observation 10: state of organic layer)

18. After water is completely removed, filter magnesium sulfate with suction on the Büchner funnel. And rinse it twice with 10mL of diethyl ether. [Note 13]

19. Transfer the resulting solution to a pre–weighed 100 mL round–bottom flask. Evaporate diethyl ether by using a rotary evaporator. When the solvent is almost evaporated, phenol will remain at the bottom of the flask. [Note 14]
Weigh the flask and calculate the recovered amount of phenol. (Observation 11: weight of phenol)

<Recovering benzoic acid>

20. Keep adding 6.0 M hydrochloric acid to the phenol in Flask **B** in small portions until the pH of the solution becomes 1 on the pH paper. When acid is added, benzoic acid will precipitate to give a white–colored solid since it is sparingly soluble in water. [Note 16] (Observation 12: formation of a white crystal)

21. Collect the solid by filtering with suction on the Büchner funnel, and wash twice with 10 mL of cold water. Allow it to dry for few minutes and spread it on a clean pre–weighed filter paper to dry in the air.

22. Weigh the filter paper when it is completely dry. Calculate the recovered amount of benzoic acid and the percent yield. Measure the melting point of the obtained benzoic acid and compare it with the literature value. (Observation 13: percent yield and m.p.)

Observation 10 :

Observation 11 :

Observation 12 :

Observation 13 :

Yield

Benzoic acid	____________ g	(________ %)
Benzyl alcohol	____________ g	(________ %)
Phenol	____________ g	(________ %)

Checked by ______________________

Note

1. This experiment is to extract a desired compound from the mixture sample. Since the purpose of this experiment is to introduce the acid–base reactions in the organic compounds, the pK_a value of each compound must be learned before the experiment to understand the procedures properly.
   The three known samples here are chosen because their TLC spots can be observed under the UV lamp (254 nm) and their pK_a values differ greatly.

2. As a rule of thumb for the extraction techniques, it is strongly recommended **to keep both the organic and the aqueous layers** until you are certain which one contains the desired material.

3. As the concentration of the compound used in the experiment is high, the spots of benzoic acid and phenol will be hard to see on the TLC plate because of the tailing. Therefore, you should take a small amount of the compound made up of 3 samples, dilute it with ethyl acetate, and carry out TLC using this diluted solution.

4. Benzoic acid with the smallest pK_a value reacts first with sodium bicarbonate ($NaHCO_3$) – a weak base – and moves to the aqueous layer.

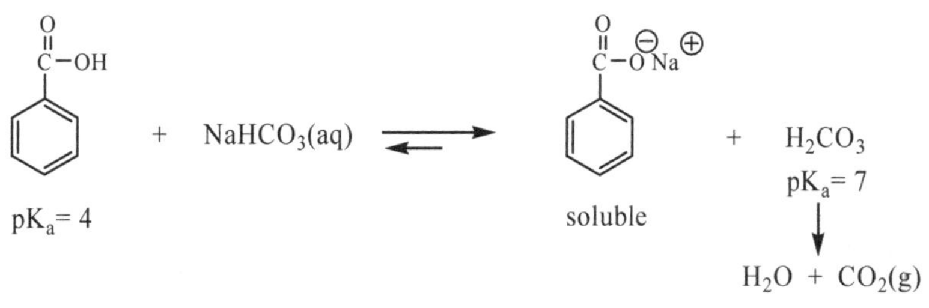

Saturated sodium bicarbonate solution is one of the **reaction solvents** which are used to remove traces of acid that are remaining in the organic layer. Likewise, an organic or mineral base present in the organic layer as impurities can be removed by washing with a dilute acid.
The usual procedure in most laboratory textbooks is adding the reaction solvent directly to the separatory funnel. But in this

textbook, I strongly recommend **adding saturated sodium bicarbonate solution to the reaction mixture in the Erlenmeyer or the round bottom flask, mix well and then transfer it to the separatory funnel in order to prevent any unexpected rapid effervescence or heat generation.**

5. If you add saturated $NaHCO_3$ solution to the reaction mixture, the overall concentration becomes diluted which results in a slower reaction rate. Put the solid $NaHCO_3$ reagent instead. Repeat until benzoic acid is completely separated from the organic layer.

6. The separatory funnel should be of such size that it must not be more than three–fourths when filled with the solutions.

7. The separatory funnel is shaken vigorously to obtain good physical mixing of the immiscible liquids. It is necessary to release the internal pressure in the separatory funnel since they build up pressure while they are mixed.
   While shaking, it is essential to grasp the funnel with both hands, one hand at the top and the other on the stopcock, so that the stopper and stopcock are held firmly in place.

8. When the layers are separated, the lower layer is drawn off carefully along the inner wall of the Erlenmeyer flask. Wash the tip of stopcock with an appropriate solvent.

9. It is important to **transfer the organic layer through the mouth of the separatory funnel** since draining through the stopcock can lead to contamination with the liquid adhering in the stem.

10. Because phenol (pK_a= 10) is a stronger acid than benzyl alcohol (pK_a= 16), it first reacts with 6.0 M sodium hydroxide, a base. If phenol is remained in the solution, add 0.30 g of sodium hydroxide in order to remove phenol completely from the organic layer.

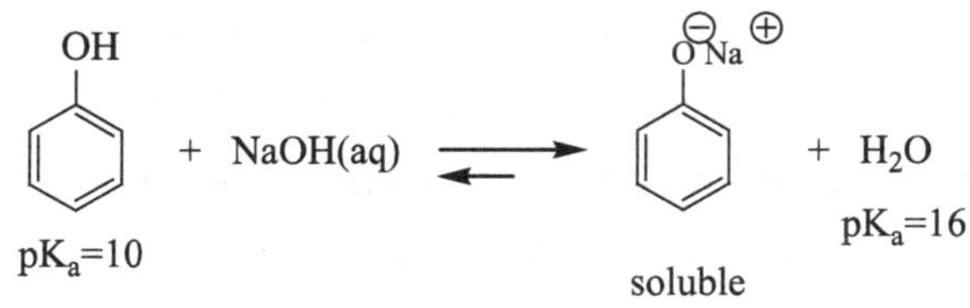

11. There could be a small amount of water remaining in the organic layer after the extraction. For example, ether which is frequently used as an extraction solvent can contain water about 1.5% of its own weight. The solubility of an organic substance in water is markedly affected by the presence of dissolved inorganic salts, especially sodium chloride. Using brine is good example of **salting-out** . (refer to **1.1 Dissolution**)

12. Treating the organic layer with brine is not enough to remove traces of water. When the organic layer is still cloudy, a drying

Common Drying Reagents

	Capacity	Rate	Use
Magnesium sulfate	7 H_2O	fast	used for general purpose. stable and fast but weakly acidic.
Sodium sulfate	10 H_2O	slow	used for acid sensitive compounds but slow.
Calcium sulfate (Drierite)	1/2 H_2O	very fast	used for drying tube and as the moisture indicator with cobalt chloride coated (blue).
Calcium chloride	6 H_2O	very fast	used for hydrocarbons or alkyl halide but may react with nitrogen or oxygen in the compound.
Molecular sieve	high	fast	used for keeping the organic solvents under drying condition. The combination of 3A and 4A is effective.
Sodium hydroxide	very high	fast	used for the amines. Powerful but strongly basic.
Sulfuric acid	very high	very fast	used for saturated of aromatic hydrocarbons. Powerful but strongly acidic.

agent is added. In the case of anhydrous magnesium sulfate, at first it sticks on the bottom of the flask to form a hydrate complex. Add small portions at a time until it moves freely in the solution and the resulting solution becomes clear.

13. Diethyl ether is evaporated very rapidly under the suction filtration and causes a gap between the filter paper and the Büchner funnel. The solid magnesium sulfate might escape through the gap to the filter flask. Keep the filter paper moist with diethyl ether during the filtration.

14. The condensed product contains a small amount of the solvent since the solvent is not completely removed by rotary evaporation. In order to remove the solvent completely, a higher vacuum should be applied.

15. About 15.0 mL of 6.0 M hydrochloric acid will be required until pH=1. As hydrochloric acid is a stronger acid than phenol, the reaction completely goes forward.

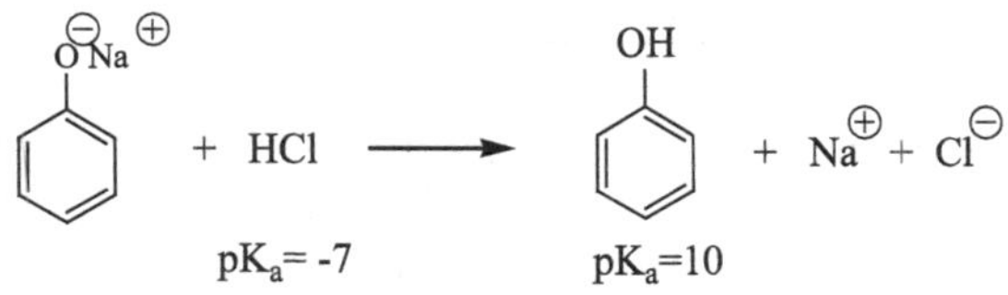

16. About 8.0 mL of 6.0 M hydrochloric acid will be required until pH=1. As hydrochloric acid is a stronger acid than benzoic acid, the reaction completely goes forward. And the resulting benzoic acid is removed from the reaction mixture as a precipitate.

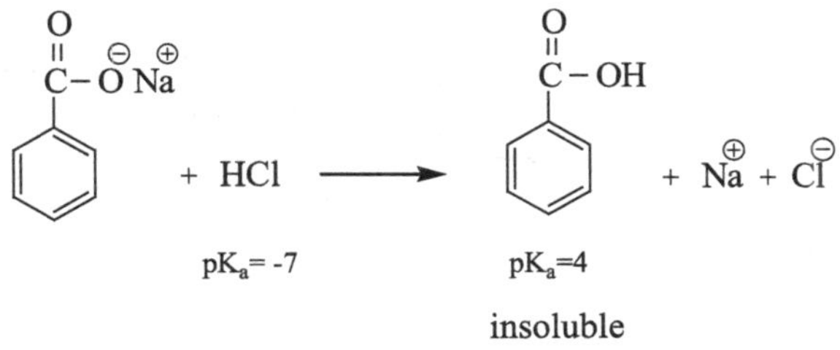

Keywords for Extraction

Summary for pK_a

Overall Scheme

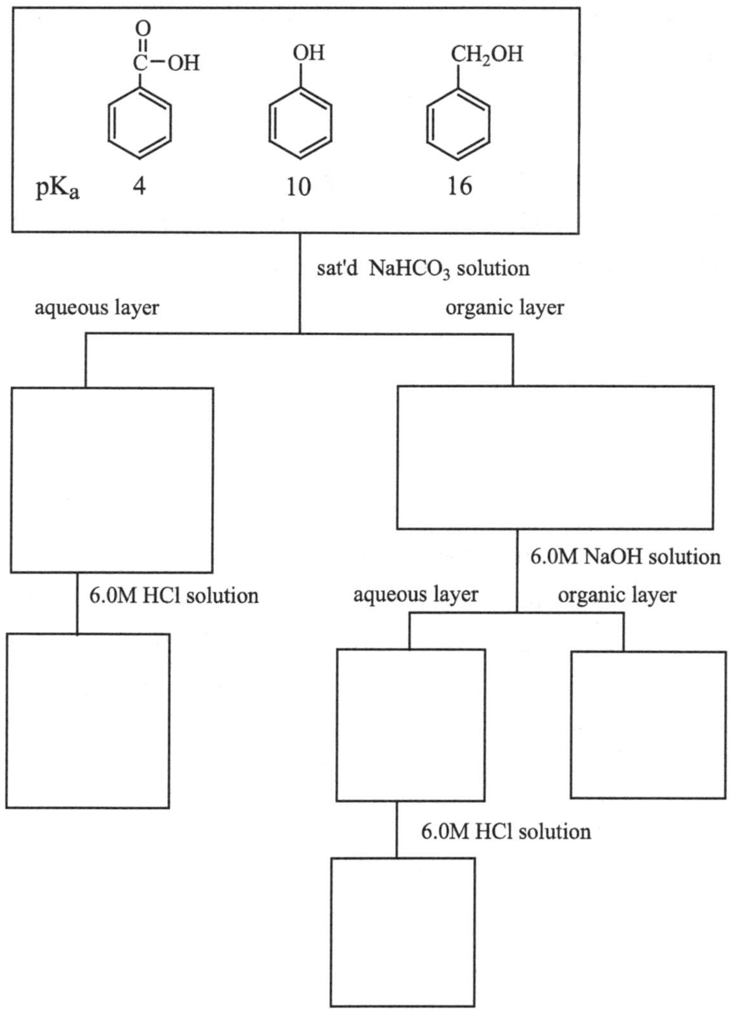

1.5 Recrystallization and Melting point

Keywords : solubility, mother liquor, freezing or melting point, melting point range, packing

One of the most significant examples of **recrystallization** in our daily life is the production of salt from the salt pan. We get salt from sea water by using natural energy such as sunlight and wind to evaporate water. Repetition of recrystallization increases the purity of the salt. In the laboratory, recrystallization is used to separate and purify a solid substance. Currently, the technique of recrystallizing a small amount of various substances has become very crucial since X–ray crystallography is widely used to elucidate molecular structures including proteins.

Recrystallization is based on **the solubility difference between a hot solvent and a cold solvent**. When no more solid dissolves at a given temperature, the resulting solution is said to be **saturated**. At this point, the solid and the solvent are in equilibrium. By lowering the temperature, some of the solid will precipitate or crystallize in order to adapt to the new equilibrium condition.

The selection of a **suitable solvent** for recrystallization is a critical factor in recrystallization because the solubility of a substance should be high in the hot solvent and low in the cold solvent. There are several criteria for selecting a solvent. But the selection of an appropriate solvent should be based on the trial and error process. A small quantity of the material is tested with a variety of solvents or solvent pairs. In general, a low–boiling solvent is preferred for recrystallization since the solvent adhered to the crystals should be readily removed by evaporation.

The following methods facilitate recrystallization.

1. Co–solvents

The mixture of two solvents which are soluble in each other but differ in their capacities to dissolve the compound to be crystallized is effective for recrystallization. For example, solvent pairs such as acetone–water, ethanol–water, diethyl ether–petroleum ether and acetic acid–water are used.

2. Lowering the temperature

When a substance has a high solubility difference in a chosen solvent according to the temperature, it can be precipitated out by cooling the hot saturated solution.

3. Evaporation

When a substance is very soluble in a chosen solvent, evaporating off a portion of the solvent may induce crystallization.

During the process of a crystal formation, the molecule tends to deposit to a growing crystal of the same molecules because it fits better in the crystal lattice for the same molecule than for the others. If the crystallization process occurs under near–equilibrium conditions, the preference to deposit on the surfaces of the same molecules will lead to an increase in the purity of the crystalline substance.

The rate of crystal formation is also an important factor in obtaining a pure crystal. Too rapid crystal growth induced by sudden cooling will not allow equilibrium conditions and produce very small crystals that might adsorb impurities on their surfaces. At the other extreme, too slow crystal growth can result in the formation of large crystals that will contain the solvent and impurities within the crystal lattice. Ideally, crystals should be formed over a period of hours rather than seconds or months. This can generally be accomplished by allowing the solution to cool down at room temperature on the bench top, followed by chilling in an ice bath. The crystallized substance may then be isolated by filtration. The purity of the resulting crystal is readily verified by its **melting point** and its **range**. The remaining solution after the crystals have deposited is called the **mother liquor**.

The **melting point** is a basic and standard method to measure the purity and to identify a solid substance since it is simple, practical and less expensive. In addition, only a small amount of the sample is required to measure the melting point.

The melting point of a pure compound is usually defined as the temperature at which the solid exists in equilibrium with its liquid form. In general, capillary melting points used in the organic laboratory are slightly higher than the actual melting points. The reverse process of melting–liquid to solid– is called **freezing**.

Melting point is strongly affected by the ability of packing in the lattice of molecules, polarity and polarizability.

Melting points of fatty acids

Compound	Number of Carbons	Structure	Melting Point (℃)
a. Palmitic acid	16	$CH_3(CH_2)_{14}COOH$	63℃
b. Stearic acid	18	$CH_3(CH_2)_{16}COOH$	70℃
c. Arachidic acid	20	$CH_3(CH_2)_{18}COOH$	75℃
d. Oleic acid(*cis*)	18	$CH_3(CH_2)_7CH{=}CH(CH_2)_7COOH$	16℃
e. Linoleic acid(*cis, cis*)	18	$CH_3(CH_2)_4CH{=}CHCH_2CH{=}CH(CH_2)_7COOH$	-5℃

The fatty acids generally contain an even number of carbon atoms between 12 and 20. If double bonds are present, they usually have the Z (*cis*) geometry.

1. As the number of carbon increases from the palmitic acid **a** (63℃) to the arachidic acid **c** (75℃), the melting points become higher because of their large polarizability. Note that **saturated fatty acids** are in the solid form at room temperature.
   (refer to **1.3 Distillation**)

2. The melting points of long–chain fatty acids decrease with the introduction of double bonds into the carbon skeleton. The oleic acid **d** and the stearic acid **b** have the same number of carbons (18). But the melting point of **d** (16℃) is much lower than that of **b** (70℃) since the double bond causes bending or twisting of the chain to prevent the molecules from stacking to be crystallized. As unsaturation increases in the linoleic acid **e**, the melting point becomes even lower (–5℃). They are called **unsaturated fatty acids** and found as constituents of oils from the seeds of plants, such as sesame oil and peanut oil.

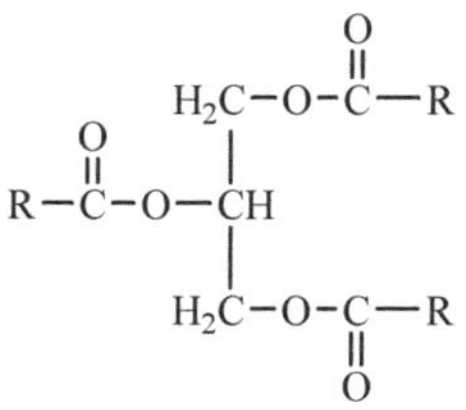

triglyceride

Fats and **oils** are esters of glycerol and fatty acids, which are called **glycerides** or **acylglycerols**. Fats are mostly made up of saturated fatty acids and usually obtained from animal sources. Oils are primarily composed of unsaturated fatty acids and commonly obtained from vegetable sources. In general, fats are solid and oils are liquid at room temperature.

Melting points of Isomers

a) Regio isomer

m.p. : 69~71℃ m.p. : 146℃ m.p. : 147~148℃ m.p. : 242.4℃

In general, a *para* isomer in an aromatic compound tends to fit the lattice more readily than an *ortho* or a *meta* isomer in aromatic compounds.

b) Geometric isomer

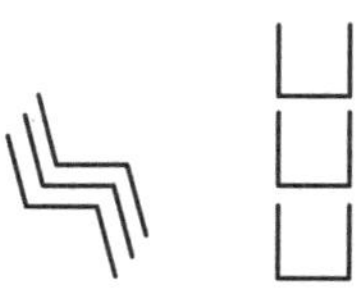

trans cis

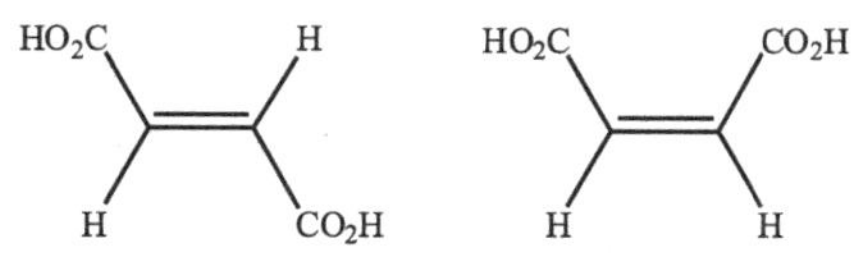

m.p. : 287℃ m.p. : 131℃

Fumaric acid Maleic acid

The geometry of the *trans* isomer is more easily packed than that of the *cis* isomer. The fumaric acid shows a much higher melting point than the maleic acid.

Because the mixture lowers the melting point of each component, sodium chloride and/or calcium chloride are often spread on streets to prevent from ice forming in freezing weather. This is a good example of **freezing–point depression.**

In this experiment, the basic technique of recrystallization to separate a solid compound and to measure the melting point of the substance are introduced.

Experiment I

Reagents

Name	Molecular formula	Structural formula	M. W.	m.p. (℃)	b.p. (℃)	Density (g/mL)
NaCl						
Dark sugar						

Apparatus

⇨ magnetic stirrer, magnetic spin bar, lab jack, alcohol lamp

Glassware

⇨ Erlenmeyer flask, graduated cylinder, filter flask, Büchner funnel, thermometer

Setting the experiment equipment

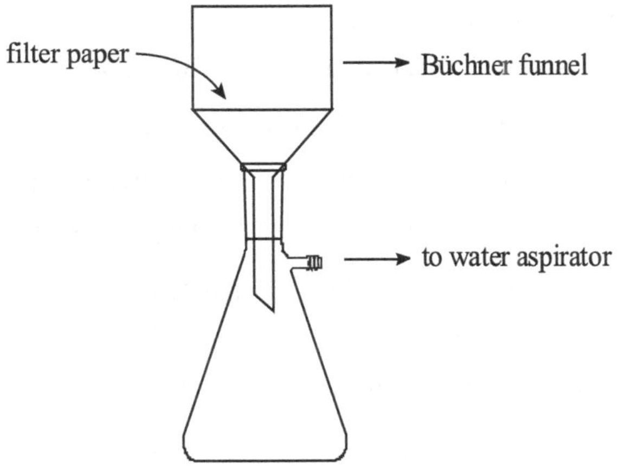

Experimental procedure

1. In a 100 mL Erlenmeyer flask, place 6.00 g of sodium chloride and 1.00 g of dark sugar. Add 20 mL of water and dissolve the mixture. [Note 1,2] (Observation 1: exact amount of reagent and solvent)
2. After the dissolution is complete, add acetone by small portions about 5.0 mL each until no further solid is precipitated. [Note 3] (Observation 2: total amount of acetone and state of precipitate)
3. For the suction filtration, place a filter paper in the Büchner funnel and start a gentle vacuum by a water aspirator. Wet the filter paper with plenty of acetone. Make sure the filter paper is sealed tightly in place.
4. Pour the resulting solution into the prepared Büchner funnel under reduced pressure. After the solid has been collected, wash it with a small amount of cold acetone.
5. **Disconnect the aspirator hose before turning off the water aspirator**. The vacuum should always be released before the water flow is turned down.
   (Observation 3: state of the collected solid)
6. Allow all of the residual solvent to evaporate from the solid. Most of the solvent can be removed by squeezing the solid between sheets of a filter paper before allowing them to air dry.
7. Sodium chloride is verified by the Beilstein Flame test.
   Take a small amount of the collected solid at the tip of the spatula and ignite in the flame to observe the color of the flame.

   [Note 4] (Observation 4: color of the flame)

Date of experiment: **Room temperature:**

Observation

Observation 1 :

Observation 2 :

Observation 3 :

Observation 4 :

- Theoretical yield ____________ g
- Practical yield ____________ g
- Percent yield ____________ %

Checked by ______________________

Note

1. The solubilities (solubility – g of solute per 1.0 mL of solvent) of salt and sugar are as the following.

Solvent 1.0mL	Sucrose	NaCl
Water (25℃)	2.0g	0.357g
Acetone	Insoluble	Insoluble

As shown in the table above, salt and sugar are soluble in water which makes water (dielectric constant =78.5) the appropriate solvent. Adding a less polar solvent to water can reduce the polarity of the resulting solution and decrease the solubility of sugar and salt. When acetone (dielectric constant = 20.7) is added, salt is precipitated first since the solubility of sugar is much higher than that of salt.

2. Both sugar and salt dissolve well in water but have quite different structures. Sugar is composed of glucose and fructose and is soluble in water because of the interaction between the hydroxyl groups and water. Meanwhile, salt is also readily soluble in water since it has an ionic bond.

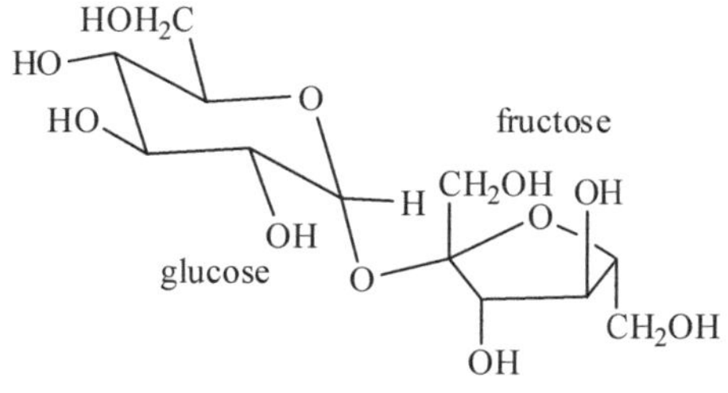

Sucrose

3. Precipitation starts when about 10.0 mL of acetone is added.

Color	Atom
Red	Li, Ca
Yellow	Na
Green	Co, Ba
Blue	Pb, Se
Violet	K

4. The Beilstein's flame test is an old, but simple and reliable method to detect the elements in the substance. You can observe the color of the flame while heating the test sample with Bunsen burner or alcohol lamp. If a yellow flame is observed, the sample may contain sodium. When sugar is heated, it boils and turns black. A greenish flame indicates that the compound contains a halogen element.

Experiment 2

Reagents

Name	Molecular formula	Structural formula	M. W.	m.p. (℃)	b.p. (℃)	Density (g/mL)
o-Nitro benzoic acid						
p-Nitro benzoic acid						

Apparatus

⇨ magnetic stirrer, magnetic spin bar, lab jack, Melting point meter

Glassware

⇨ Erlenmeyer flask, graduated cylinder, filter flask, Büchner funnel, thermometer

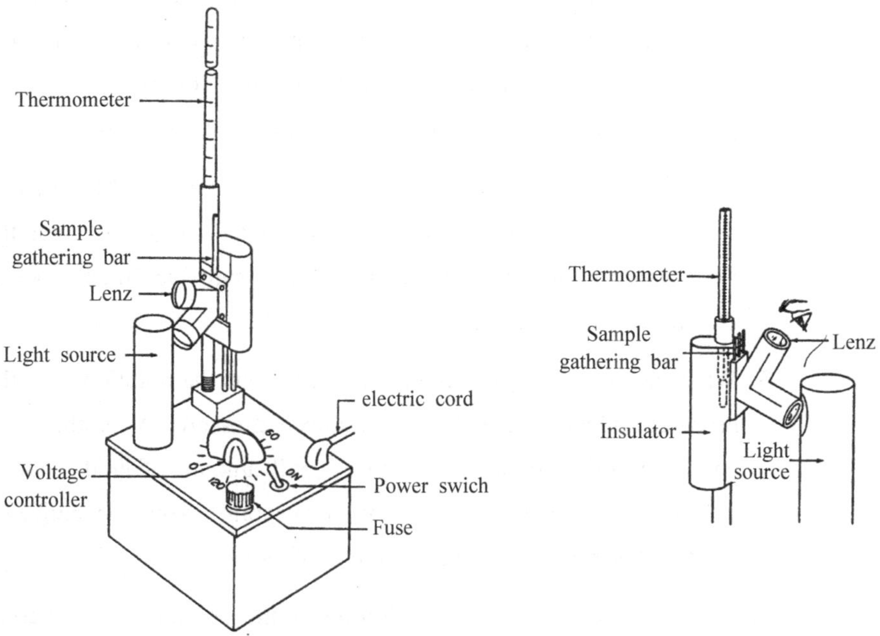

Mel–Temp Apparatus

Experimental procedure

1. Mix 2.00 g of *p*–nitrobenzoic acid and 0.40 g of *o*–nitrobenzoic acid in a 50 mL Erlenmeyer flask and prepare two capillary tube samples to check the melting point. Add 25.0 mL of ethanol and apply heat until it dissolves completely. [Note1, 2]
   (Observation 1: exact amount of reagent and solvent)
2. After the mixture is dissolved completely, cool it at room temperature on the bench until crystallization happens. Then put the reaction flask in a ice bath for further crystallization.
   (Observation 2: formation statues and time)
3. During the crystal formation, check the melting points of *p*–nitrobenzoic acid, *o*–nitrobenzoic acid and the mixture using an appropriate melting point apparatus. In this textbook, the commercial Mel–Temp apparatus is used.
   It is advisable to prepare two capillary tubes for each sample. Insert a capillary tube containing a sample in the slot in front of the thermometer. The tube is then heated at a controlled rate and the melting point range is recorded. [Note 3 or 4]
   (Observation 3: melting point range and melting point)
4. For the suction filtration, place a filter paper in the Büchner funnel and start a gentle vacuum by a water aspirator. Wet the filter paper with plenty of ethanol. Make sure the filter paper is sealed tightly in place.
5. When no further crystal is precipitated, pour the solution into the prepared Büchner funnel under reduced pressure. After the solid has been collected, wash it with a small amount of cold ethanol. Allow all of the residual solvent to evaporate from the collected solid. (Observation 4: used amount of ethanol)
6. To complete the filtration, **disconnect the aspirator hose before turning off the water aspirator**. Vacuum should always be released before the water flow is turned down.
   Keep the mother liquor for the TLC check.

Date of experiment : **Room temperature:**

Observation

Observation 1 :

Observation 2 :

Observation 3 :

Observation 4 :

7. Most of the solvent can be removed by squeezing the solid between sheets of filter paper before allowing them to air dry until the next laboratory period.

8. Check and compare the TLC of the obtained solid and the mother liquor. <the developing solvent is ethyl acetate: hexane=1:1. To prevent tailing, add one or two drops of formic acid in the developing solvent.> (Observation 5: TLC)

9. Weigh and check the melting point of the isolated solid and compare it with the data obtained in procedure 3.
(Observation 6: melting points of the samples)

Observation 5 :

Observation 6 :

- Theoretical yield ____________ g
- Practical yield ____________ g
- Percent yield ____________ %

Checked by ______________________

Note

1. The typical procedure for recrystallization is to prepare a hot solution and add it to the solid sample until it completely dissolves. But in this textbook, we first add the solvent to the solid mixture and then heat it up to dissolve completely since the preparation and handling of a hot solution might be dangerous.

2. The solubilities of nitrobenzoic acids are as the following.

Solvent 1.0mL	*o*-nitrobenzoic acid	*p*-nitrobenzoic acid
Water	6.84 mg	0.42 mg
Ethanol	340 mg	9.09 mg
Ethanol (boiling)	Soluble	91.0 mg

As shown in the table, a *para* isomer can easily be made into a saturated solution in ethanol at room temperature because of its low solubility. Ethanol is chosen as the appropriate solvent for the recrystallization.

3. Care must be taken with heating. It should not be done too rapidly otherwise the results will not be consistent. The heating rate of 2℃ per minute is suitable for the common organic compounds. However, since heating at this rate from room temperature to 250℃ would take an excessive amount of time, preparation of two samples is strongly recommended. The first one should be heated rapidly to get a rough idea of its melting point. The second sample is then heated rapidly to approximately 20℃ below where the first sample started melting and then controlled heating rate will be used beyond that point.

4. A **melting point range** is the interval between the temperature at which the compound begins to liquefy and where it becomes a clear liquid. When the solid substance is pure, the melting point range is about 1–2℃. But the presence of a small amount of impurities will lower the melting point and broaden the melting point range. These facts are widely applied in establishing the identities of organic compounds by determining the melting points of the mixtures – **mixed melting point**. The purer the compound, the narrower the melting point range it will have.

Experiment 3

Reagents

Name	Molecular formula	Structural formula	M. W.	m.p. (℃)	b.p. (℃)	Density (g/mL)
o-Nitroaniline						
p-Nitroaniline						

Apparatus

⇨ magnetic stirrer, magnetic spin bar, lab jack, Melting point meter

Glassware

⇨ Erlenmeyer flask, graduated cylinder, filter flask, Büchner funnel, thermometer

Experimental procedure

1. Mix 1.00 g of *p*–nitroaniline and 0.20 g of *o*–nitroaniline in a 50 mL Erlenmeyer flask and prepare two capillary tube samples for measuring the melting point. Add 7.0 mL of ethanol and apply heat until it dissolves completely. [Note 1, 2]
   (Observation 1: exact amount of reagent and solvent)

2. After the mixture is dissolved completely, cool it at room temperature on the bench until crystallization happens. Then put the reaction flask in the ice bath for further crystallization.
   (Observation 2: formation statues and time)

3. During the crystal formation, check the melting points of *p*–nitroaniline, *o*–nitroaniline and the mixture using an appropriate melting point apparatus. In this textbook, the commercial Mel–Temp apparatus is used.
   It is advisable to prepare two capillary tubes for each sample. Insert a capillary tube containing a sample in the slot in front of the thermometer. The tube is then heated at a controlled rate and the melting point range is recorded. [Note 3 or 4]
   (Observation 3: melting point range and melting point)

4. For the suction filtration, place a filter paper in the Büchner funnel and start a gentle vacuum by a water aspirator. Wet the filter paper with plenty of ethanol. Make sure the filter paper is sealed tightly in place.

5. When no further crystal is precipitated, pour the solution into the prepared Büchner funnel undera reduced pressure. After the solid has been collected, wash it with a small amount of cold ethanol. Allow all of the residual solvent to evaporate from the collected solid.[Note 5] (Observation 4: used amount of ethanol)

6. To complete the filtration, **disconnect the aspirator hose before turning off the water aspirator**. The vacuum should always be released before the water flow is turned down.
   Keep the mother liquor for TLC check.

Date of experiment : **Room temperature:**

Observation

Observation 1 :

Observation 2 :

Observation 3 :

Observation 4 :

7. Most of the solvent can be removed by squeezing the solid between sheets of filter paper before allowing them to air dry until the next laboratory period.
8. Check and compare the TLC of the obtained solid and the mother liquid. <the developing solvent is ethyl acetate: hexane=1:1. To prevent tailing, add one or two drops of formic acid in the developing solvent.> [Note 5] (Observation 5: TLC)
9. Weigh and check the melting point of the isolated solid and compare it with the data obtained in procedure 3. [Note 6,7] (Observation 6: melting point of the samples)

Observation 5 :

Observation 6 :

- Theoretical yield ____________ g
- Practical yield ____________ g
- Percent yield ____________ %

Checked by ______________________

Note

1. The typical procedure for recrystallization is to prepare a hot solution and add it to the solid sample until it completely dissolves. But in this textbook, we first add the solvent to the solid mixture and then heat it up to dissolve completely since the preparation and handling of a hot solution might be dangerous.

2. The solubility of *o*, *p*–nitroanilines are as the following.

Solvent 1 mL	*o*–nitroaniline	*p*–nitroaniline
Water	6.84 mg	0.42 mg
Ethanol	229 mg	40.0 mg
Ethanol (boiling)	2230 mg	163 mg

As shown in the chart, a *para* isomer can easily be made into a saturated solution in ethanol at room temperature because of its low solubility. Ethanol is chosen as the appropriate solvent for recrystallization.

3. Care must be taken with heating. It should not be done too rapidly otherwise the results will not be consistent. The heating rate of 2℃ per minute is suitable for the common organic compounds. However, since heating at this rate from room temperature to 250℃ would take an excessive amount of time, preparation of two samples is strongly recommended. The first one should be heated rapidly to get a rough idea of its melting point. The second sample is then heated rapidly to approximately 20℃ below where the first sample started melting and then controlled heating rate will be used beyond that point.

4. A **melting point range** is the interval between the temperature at which the compound begins to liquefy and where it becomes a clear liquid. When the solid substance is pure, the melting point range is about 1–2℃. But the presence of a small amount of impurities will lower the melting point and broaden the melting point range. These facts are widely applied in establishing the identities of organic compounds by determining the melting points of the mixtures – **mixed melting point**. The purer the compound, the narrower the

melting point range it will have.

5. *o*, *p*–Nitroaniline can be verified according to the shape and color of the crystallization. After recrystallization, only the *para* compound can be crystallized.

	o–Nitroaniline	*p*–Nitroaniline
Color of Crystall	Orange	Greenish yellow
Shape of Crystall	The shape of star	The shape of plane

6. As discussed in **1.2 Thin Layer Chromatography**, the *para* isomer of nitroaniline holds a bigger dipole moment than does the *ortho* isomer because of the resonance contribution that creates a large distance between the two charges.

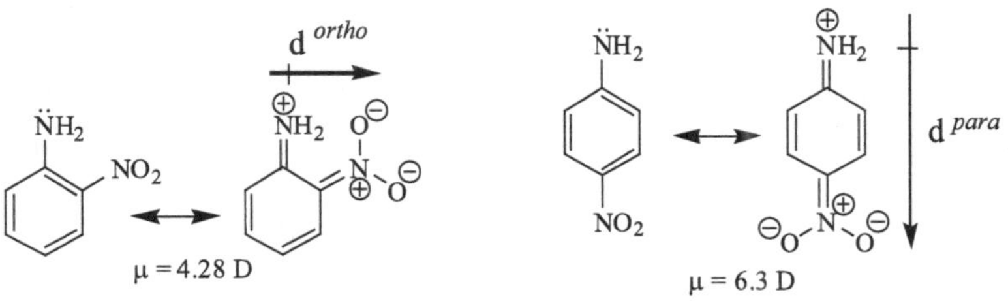

$d^{ortho} < d^{para}$

Therefore, *p*–nitroaniline shows a lower R_f value than *o*–nitroaniline on TLC. The R_f value of *o*–nitroaniline is 0.62 and the R_f value of *p*–nitroaniline is 0.44 (ethyl acetate : hexane = 1 : 2)
m.p. of *p*–nitroaniline after recrystallization : 140.0 ~ 142.0℃
(Lit. value : m.p. of *p*–nitroaniline is 149.0 ~ 151℃)

7. When the crystal growth is relatively slow and selective, it is called **crystallization.** On the contrary, if the process is rapid and nonselective, it is called **precipitation**.

Keywords for Recrystallization

Summary for Melting point

CHAPTER

02

Basic Organic Reactions

2.1 Esterification

Keywords : wax, fat, oil, lipid, pheromone, lactone, protection, Le Châtelier's principle

Fats and **waxes** are esters of long–chain fatty acids and alcohols.

Waxes contain long–chain primary alcohols such as the cetyl alcohol. They are used for protecting the surfaces – as in shoe polish and car wax – and also used in pharmacy to harden soft ointments. The surfaces of leaves and fruits are coated with a waterproof layer of wax.

$CH_3(CH_2)_{14}CH_2OH$
cetyl alcohol
(1-hexadecanol)

Animal fats and **vegetable oils** are triacyglycerols (triglycerides), esters of glycerol and various fatty acids. The acids found in fats are mostly saturated, while those in oils tend to be unsaturated. In general, fats are solid and oils are liquid at room temperature since unsaturation lowers the melting point.

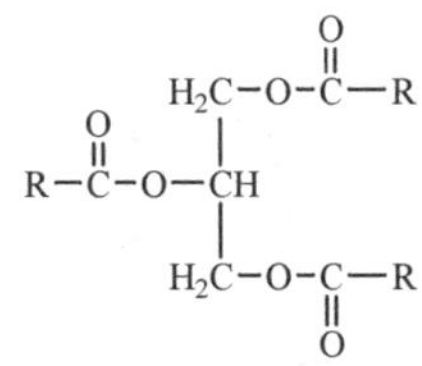

triglyceride

Lipids are naturally occurring organic molecules, which have limited solubility in water. The typical lipids include fats, waxes and oils.

The simple esters give the taste and flavor of flowers and/or fruits. For example, benzyl acetate gives the peach scent and octyl acetate gives the orange scent. Especially, isoamyl acetate is the ingredient of both banana oil and honeybee's alarm pheromone. **Pheromones** are secreted by insects and draw characteristic physiological and behavioral responses from other members of the same species.

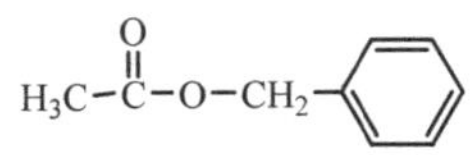

Benzyl acetate (peach)

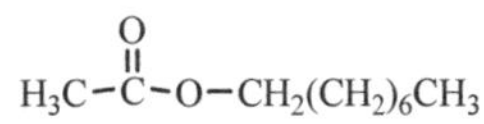

Octyl acetate (orange)

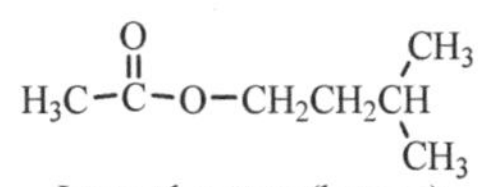

Isoamyl acetate (banana)
alarm pheromone

A polymer connected by the ester linkage is called polyester. [Poly (ethylene terephthalate)], also known as the trade name Dacron, Mylar or Terylene, is widely used in fibers, recording tapes, X–ray films and tire cords due to its good chemical and solvent resistance. The **ester interchange** is often the most practical reaction for polyester synthesis. Since diacid has poor solubility and is not easily purified, methyl esters are used instead. Polymerization is driven to the completion by removal of alcohol from the reaction mixture.

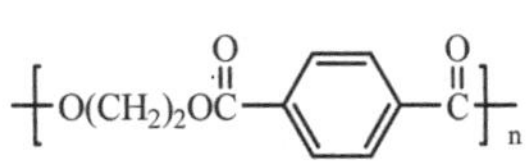

Poly(ethylene terephthalate)
(P.E.T)

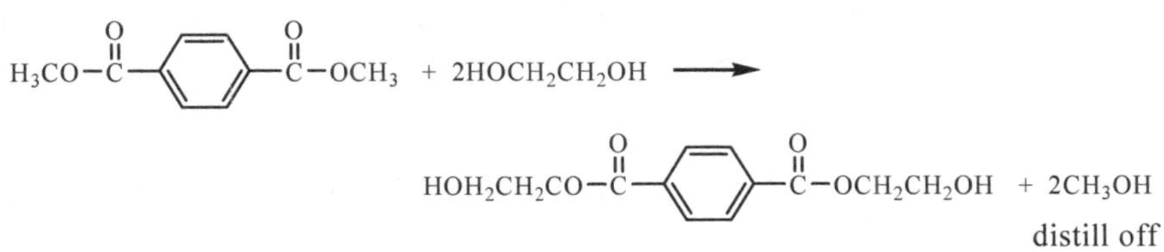

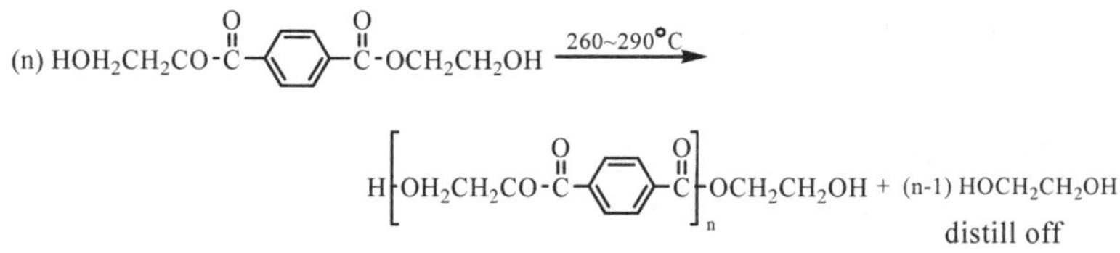

When esterification occurs within a molecule, **lactone** is obtained.

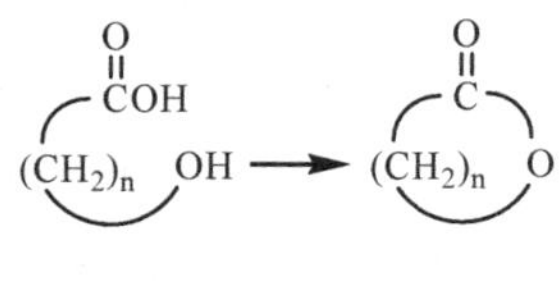

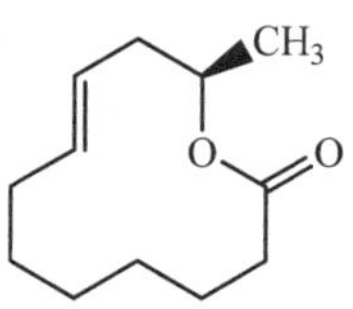

Recifeiolide
(antibiotics)

Large–ring lactones isolated from natural sources have various **antibiotic** and **antitumor** activities and are called **macrolides** (for example, recifeiolide). They contain 12 to 17 carbon atoms in a cyclic skeleton.

During the multi–step organic synthesis, it is necessary to protect the carboxyl group by esterification because of its acidity. Several useful protection groups are shown below.

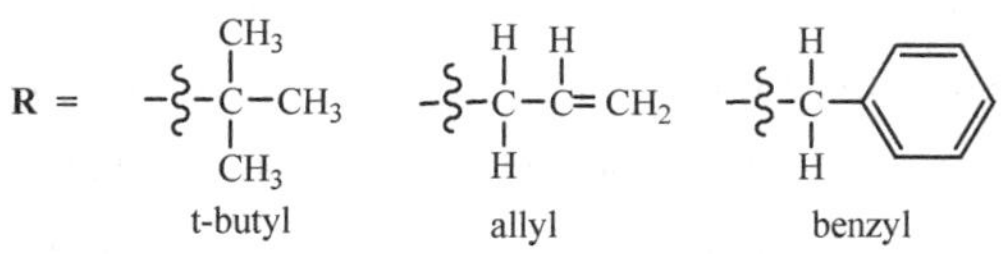

Following reagents are developed in order to facilitate esterification under a mild reaction condition and **coupling reagents**. They convert the hydroxyl group of the carboxyl group into a better leaving group.

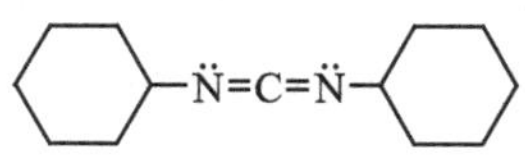

DCC
(1,3-Dicyclohexylcarbodiimide)

$EtO_2C\ddot{N}=\ddot{N}CO_2Et$

DEAD
(Diethylazodicarboxylate)

Diazomethane is applied to make methyl ester in a single step.

$$RCOOH + CH_2N_2 \longrightarrow RCOOCH_3 + N_2$$

Methyl ester is one of the functional groups used to protect carboxylic acid *in vivo*.

In this experiment, benzoic acid reacts with methanol under an acid catalyst condition.

Chemical equation

Experiment I

Reagents

Name	Molecular formula	Structural formula	M. W.	m.p. (℃)	b.p. (℃)	Density (g/mL)
Benzoic acid						
Methyl alcohol						
Methyl benzoate (Products)						

Apparatus

⇨ magnetic stirrer, magnetic spin–bar, lab–jack, heating mantle, transformer, UV lamp, vacuum pump, evaporator

Glassware

⇨ TLC chamber, 100 mL two–neck round bottom flask, 100 mL graduated cylinder, reflux condenser
100 mL Erlenmeyer flask, 100 mL separatory funnel, filter flask, Büchner funnel

Experimental procedure

<Reaction>

1. In a 100 mL two–neck round bottom flask, place 5.00 g (40.9 mmol, 1.0 eq) of benzoic acid and 40.0 mL (31.6 g, 986.0 mmol, 24.8 eq) of methanol. [Note 1, 2]
   To monitor the reaction, prepare 3 or 4 TLC plates and spot them with the starting material.
2. Slowly add 1.1 mL (2.00 g, 20.6 mmol, 0.5 eq) of concentrated sulfuric acid, and then add the magnetic spin bar. [Note 3]
3. Assemble the reflux apparatus. Make sure that the liquid level of the reaction mixture is higher than the level of the heat source. [Note 4] (refer to **1.3 Distillation**)
   Start reflux while the mixture is stirred thoroughly.
   (Observation 1: change in temperature, time of reaction)
4. After reflux has begun, monitor the reaction by the pre–spotted TLC plate every 20 minutes and observe if the benzoic acid has disappeared. <the developing solvent is ethyl acetate: hexane = 1: 4> [Note 5] (Observation 2: TLC)
5. When the benzoic acid has completely disappeared on the TLC plate, continue reflux for 10 minutes. [Note 6] Remove the heat source and allow the reaction flask to cool down to room temperature.
   (Observation 3: TLC and reaction time)

<Work–up>

6. Disassemble the reflux apparatus.

   Add 30.0 mL of water and 35.0 mL of diethyl ether to the reaction flask. Saturate the reaction mixture by adding enough amount of solid sodium chloride and stir for 2~3 minutes. [Note 7]

   (Observation 4: state of Solution)

Date of experiment: **Room temperature:**

Observation

Observation 1 :

Observation 2 :

Observation 3 :

Observation 4 :

7. Transfer the resulting reaction mixture to the separatory funnel using a conical funnel. Allow it to stand undisturbed until the layers are completely separated. And then the lower layer— aqueous layer— is drawn off carefully through the stopcock into a 100 mL Erlenmeyer flask.

8. Wash the organic layer with 25.0 mL of water. The organic layer is poured through the mouth of the separatory funnel into a clean Erlenmeyer flask. Rinse the empty separatory funnel with 10.0 mL of diethyl ether. [Note 8]

9. To the Erlenmeyer flask containing the organic layer, add 30.0 mL of saturated $NaHCO_3$ solution and stir the mixture vigorously until no more gas is evolved. Check pH of the mixture. [Note 9] (Observation 5: state of substance and pH)

10. Transfer it to the separatory funnel and then remove the aqueous layer. The organic layer is washed with brine and then poured through the mouth of the separatory funnel into a clean Erlenmeyer flask. Rinse the empty separatory funnel with 10.0 mL of diethyl ether. Add a small amount of anhydrous magnesium sulfate ($MgSO_4$) to the organic layer in an Erlenmeyer flask until the solution becomes clear and the solid $MgSO_4$ moves freely in the solution

11. Filter the resulting organic layer in the appropriate suction filtration procedure, which is dealt in **1.5 Recrystallization**.

12. Transfer the resulting solution (**filtrate**) to a pre-weighed 100 mL round bottom flask (**tared flask**). Evaporate diethyl ether by using a rotary evaporator. When the solvent is almost evaporated, methyl benzoate will remain at the bottom of the flask.

13. Weigh the flask and check the purity of the product by TLC. Calculate the percent yield. [Note 10]
(Observation 6: TLC and percent yield)

Observation 5:

Observation 6:

- Theoretical weight of product ____________ g
- Practical weight of product ____________ g
- Percent yield ____________ %

Checked by ______________________

Note

1. The acid catalyzed esterification is a reversible reaction.

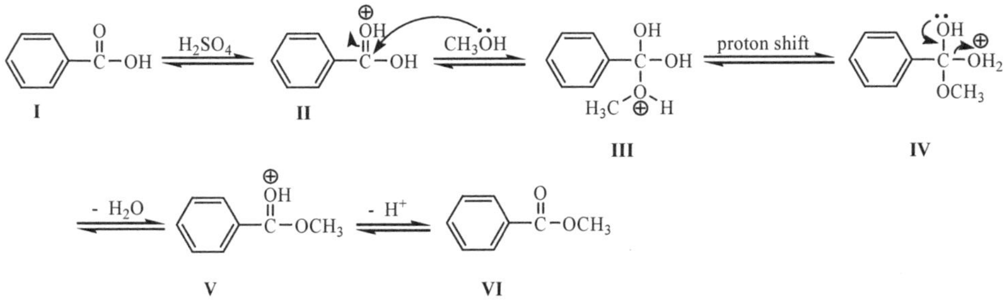

The protonation of the oxygen atom of the carbonyl group enhances the reactivity of the carboxylic acid toward the nucleophilic acyl substitution reaction. (I → II) The desired ester is obtained by removal of water followed by deprotonation. (IV → VI) The net transformation is the substitution of an –OH group by –OR.

2. To produce a good yield of ester, it is necessary to control the reaction conditions. According to Le Châtelier's Principle, several methods are available.
   1) addition of an excess amount of one of the reactants, usually the alcohol
   2) removal of ester from the reaction mixture
   3) removal of water by azeotropic distillation or by using a dehydrating agent (e.g. silica gel or a molecular sieve).

3. In the case of large–scale reactions, cool down the reaction flask in an ice bath and slowly add sulfuric acid by small portions.
   1.00 g of sulfuric acid is about 33 drops by the common disposable pipette.

4. During reflux of the reaction mixture, the boiling mixture should be confined to the reaction flask while the hot vapor enters into the cooled condenser where it condenses. **The height of the condensed vapor should not be over one quarter of the water–cooled condenser.**

5. It is difficult to take the sample using a capillary tube during the reflux because of the hot vapor from the solvent. When checking the reaction mixture, cool down the reaction flask by removing the heat source and then take the TLC. R_f of benzoic acid : 0.25 and R_f of methyl benzoate (product) = 0.75

6. After 40 minutes, the decreased size of the spot of benzoic acid can be observed. If benzoic acid is still observed on the TLC plate after one hour, add 10.0 mL (7.90 g, 247.0 mmol, 6.0 eq) of methanol and reflux for another 20 minutes. Check the disappearance of benzoic acid again.

7. Sometimes, the reaction mixture may not separate clearly into the aqueous layer and organic layer since methanol has similar polarity with water. In this occasion, add enough amount of solid sodium chloride until the layers are separated. **(Salting-out effect)**

8. The reason for washing with water is to remove the remaining sulfuric acid from the organic layer. When extracting, shake the separatory funnel 2~3 times and make sure the gas is completely removed.

9. Saturated sodium bicarbonate solution is used to remove the remaining acid from the organic layer. The evolution of CO_2 gas may damage the separatory funnel and cause an accident. Therefore, stir it in the Erlenmeyer flask until gas is no longer evolved and then transfer the reaction mixture to the separatory funnel. The pH of the reaction mixture should be around 8~10. If not, repeat the same procedure again.

10. The percent yield is 63~81%.
    The yield obtained by the evaporated product is higher since diethyl ether is not completely removed by a rotary evaporation.

Keywords for Esters

Chemical equation for Synthesis of Methyl benzoate

Reaction mechanism for Synthesis of Methyl benzoate

2.2 Saponification

Keywords : fatty acid, soap, detergent, phase transfer catalyst, driving force, pK_a

Hydrolysis of animal fats or vegetable oils using a base gives rise to the salts of long chain carboxylic acids (soap) and an alchohol (glycerol). This reaction is called **saponification** because it is used for making **soap**.

A long–chain acid salt has two domains: the **hydrophilic** head (the ionic carboxylate group that is soluble in water) and the **hydrophobic** tail (the long hydrocarbon portion that dissolves the organic residues including hydrocarbons and/or aromatic compounds.) Molecules containing hydrophilic heads and long–chain hydrophobic tails play an important role in the structure of **membranes** in a cell.

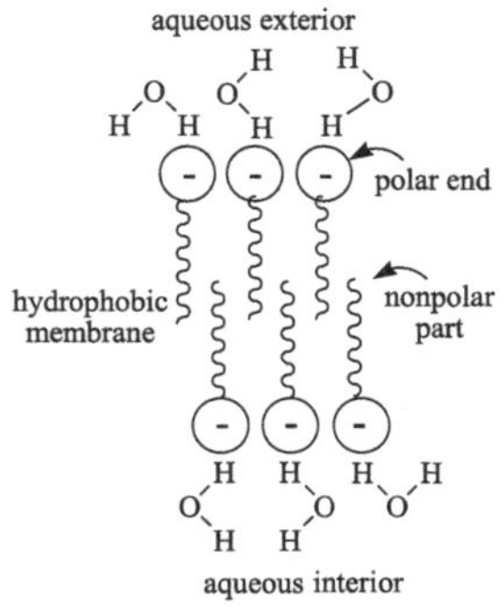

Cell membrane

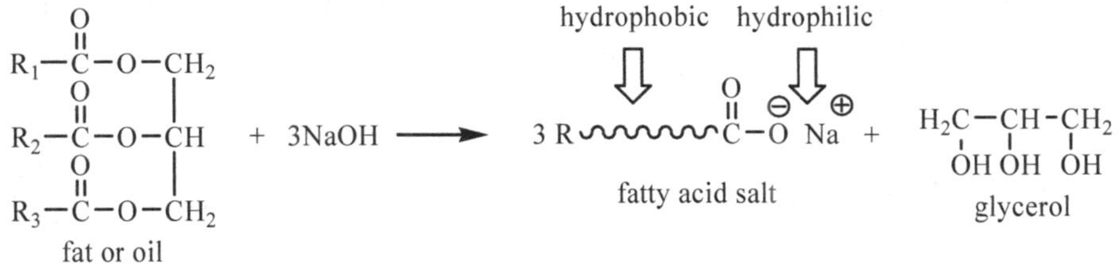

The length of hydrocarbon and the number of double bonds in the fat or oil determine the properties of the resulting soap. The carboxylic acid salts in soaps usually contain 12–18 carbons arranged in a straight chain. While soap has the advantage of being bio–degradable, it is ineffective in hard water, which contains salts of magnesium, calcium or iron ions. When soap is used in hard water, insoluble calcium salts of the fatty acids and other precipitates deposit and leave behind their sediments or stains.

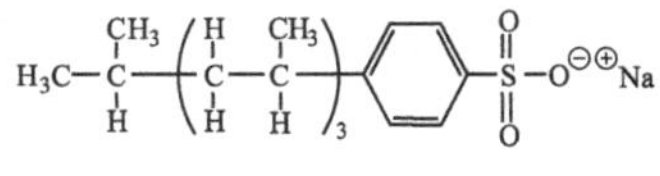

Alkyl benzene sulfonate detergent(ABS)

Detergents from inexpensive petroleum sources have been developed as an alternative to soap because they are effective in both soft and hard water. However, **alkyl benzene sulfonates** (ABS) are not biodegradable by microorganisms and can cause environmental problems. It has been found that a **linear alkyl sulfonate detergent** (LAS) containing one branch next to the aromatic ring is biodegradable by the bacterial enzyme. Now the LAS detergents replace the old non–biodegradable detergents.

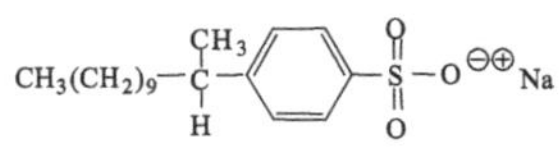

Linear alkyl sulfonate detergent(LAS)

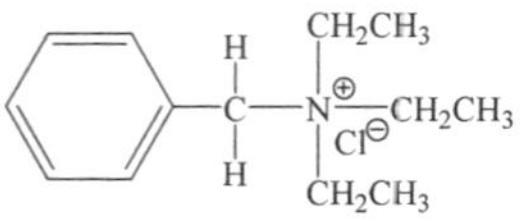

Q = Benzyl triethyl ammonium chloride

Soap, detergent, and quaternary ammonium salts hold hydrophilic and hydrophobic properties together. They are important **phase transfer catalysts**. Phase transfer catalysts are widely applied to reactions where the starting materials do not mix well under normal conditions. For example, CN^- is dissolved in the aqueous layer but not dissolved in the organic layer in a substitution reaction of CN^- to the 1–chlorooctane. When a small amount of the phase transfer catalyst–quaternary ammonium salt–is added to the aqueous layer, CN^- in the aqueous layer is transferred to the organic layer by the following mechanism.

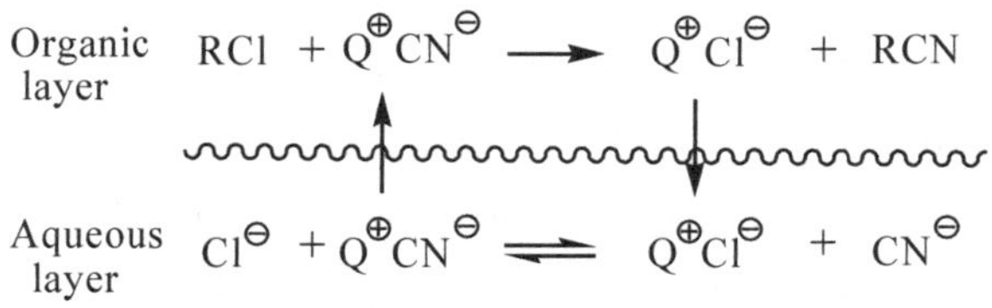

Q : quaternary salt

Saponification of *β*–keto ester followed by acidification and decarboxylation gives rise to cyclohexanone.

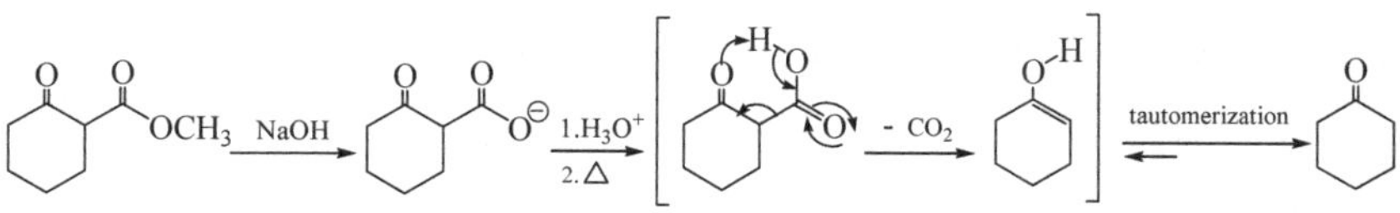

Under the condition developed by **Krapcho**, β–keto ester is converted to cyclohexanone in a one step reaction. It is reported that Cl^- is remarkably activated by the strong coordination effect of dimethyl sulfoxide (DMSO) on the cation Na^+. (refer to **2.7 Cannizzaro Note 1**) The **naked Cl^-** attacks the methyl group instead of the carbon in the carbonyl group. However, there is a disadvantage that a high reaction temperature is required.

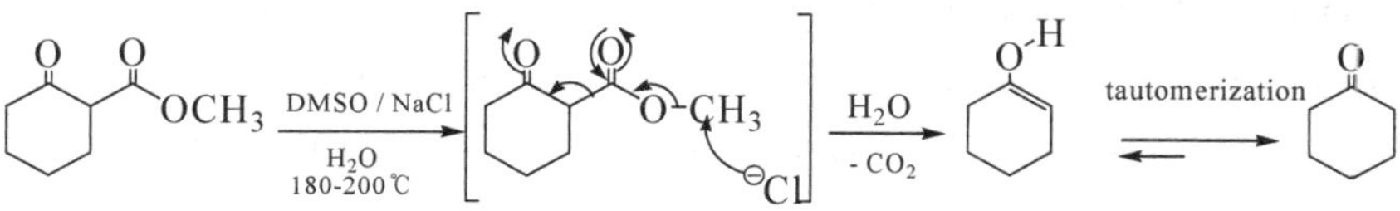

In this experiment, methyl salicylate is hydrolyzed under basic reaction condition.

Chemical equation

Experiment I

Reagents

Name	Molecular formula	Structural formula	M. W.	m.p. (℃)	b.p. (℃)	Density (g/mL)
Methyl salicylate						
Sodium hydroxide						
Salicylic acid (Products)						

Apparatus

⇨ magnetic stirrer, magnetic spin–bar, lab jack, heating mentle, transformer, melting point apparatus, UV lamp, vacuum pump, ice bath, thermometer

Glassware

⇨ TLC chamber, 100 mL two–neck round bottom flask, reflux condenser, 250 mL beaker, 100 mL Erlenmeyer flask, Büchner funnel, filter flask, watch glass

Experimental procedure

<Reaction>

1. In a 100 mL Erlenmeyer flask, place 25.0 mL of water and cool in an ice bath. Add 5.00 g (125.0 mmol, 7.5 eq) of sodium hydroxide at once. [Note 1] (Observation 1: change in temperature)
2. In a 100 mL two-neck round bottom flask, place 2.50 g (16.7 mmol, 1.0 eq) of methyl salicylate, and then slowly add the prepared sodium hydroxide solution. [Note 2, 3]
   To monitor the reaction, prepare 3 or 4 TLC plates and spot them with the starting material.
3. Add the magnetic spin bar and assemble the reflux apparatus. Make sure that the liquid level of the reaction mixture is higher than level of the heat source. Start reflux while the mixture is stirred thoroughly [Note 4]
   (Observation 2: reaction time and change in temperature)
4. Reflux the reaction mixture for 30 minutes, and check the reaction mixture by the pre–spotted TLC plate. Observe if methyl salicylate has disappeared <the developing solvent is ethyl acetate: hexane = 1:9> [Note 5] (Observation 3: change of TLC)
5. When methyl salicylate has completely disappeared on TLC, continue reflux for another 10 minutes. Remove the heat source and allow the reaction flask to cool down to room temperature. (Observation 4: reaction time)

Date of experiment : **Room temperature :**

Observation

Observation 1 :

Observation 2 :

Observation 3 :

Observation 4 :

<Work–up>

6. Disassemble the reflux apparatus.
   Transfer the reaction mixture with the magnetic spin bar to a 100 ml Erlenmeyer flask in the ice bath, and slowly add 3.0 M of sulfuric acid until its pH becomes 2 in pH paper. [Note 6]
   (Observation 5: change of solution temperature and crystal formation)
7. When no more crystal is precipitated, filter it using the appropriate suction filtration procedure that is dealt in **1.5 Recrystallization**.
   Keep a small amount of the crude product for comparison with the recrystallized product later.
   While filtering, prepare 100.0mL of boiling water for the recrystallization.
   (Observation 6: weight and color of the crude crystal)
8. Place the crude product in a 250mL Erlenmeyer flask with the magnetic spin bar, and add boiling water slowly until it dissolves completely. [Note 7] (Observation 7: amount of water, state of crystal before and after recrystallization)
9. When the solid dissolves completely, cool it down to room temperature and put it in an ice bath. The final product is obtained by the same suction filtration procedure mentioned in Procedure 7.
10. After the product is sufficiently dried, weigh the product and calculate percent yield. Check the melting point including the range and TLC of the crude product and the recrystallized product.
    [Note 8,9,10] (Observation 8: percent yield and melting point)

Observation 5 :

Observation 6 :

Observation 7 :

Observation 8 :

- Theoretical weight of product ____________ g
- Practical weight of product ____________ g
- Percent Yield ____________ %

Checked by ______________________

Note

1. Use sodium hydroxide directly after weighing it since it has deliquescent property, i.e. it readily absorbs moisture in the air.

2. A white precipitation can be observed. It is methyl salicylate which does not dissolve well because of its low solubility in water. (1.00 g/1.50 L) However, when sodium hydroxide is added to the reaction mixture, it converts methyl salicylate into a phenoxide ion, which is soluble in water.

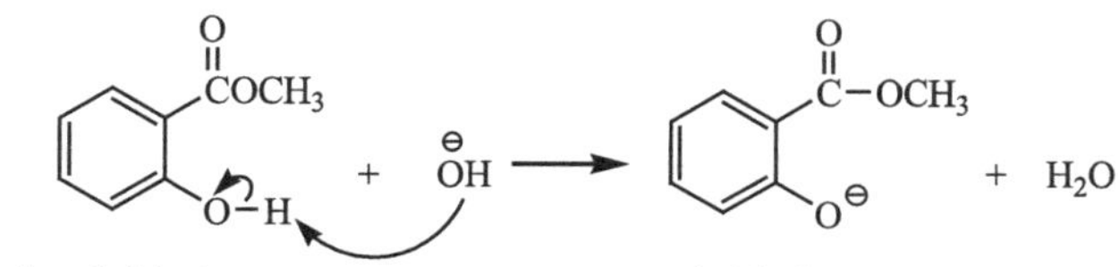

insoluble in water soluble in water

I

3. The addition of hydroxide to the compound **I** gives rise to carboxylic acid **II** and methoxide. It readily deprotonates the resulting carboxylic acid **II** to produce carboxylate **III** and methanol. The formation of a more stable carboxylate is the **driving force** for **saponification** (basic hydrolysis). The shift of equilibrium depending on pK_a is dealt thoroughly in **1.4 Extraction**.

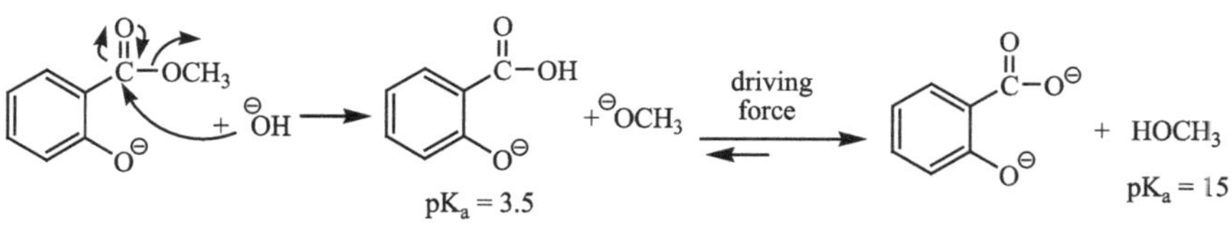

I II III

4. During reflux of the reaction mixture, the boiling mixture should be confined to the reaction flask while the hot vapor enters into the water–cooled condenser where it condenses. **The height of the condensed vapor should not be over one quarter of the water–cooled condenser.**

5. When checking TLC, the developing solvent system (acetone) is used for salicylic acid since it is not developed in the developing solvent system for methyl salicylate <the developing solvent is ethyl acetate: hexane = 1:9, R_f of methyl salicylate = 0.54>

6. The acid is recovered by the addition of a strong acid such as hydrochloric or sulfuric acid. Around 15.0 mL of 3.0 M sulfuric acid will be used. White crystal will be formed and the mother liquor will show a color of light pink.

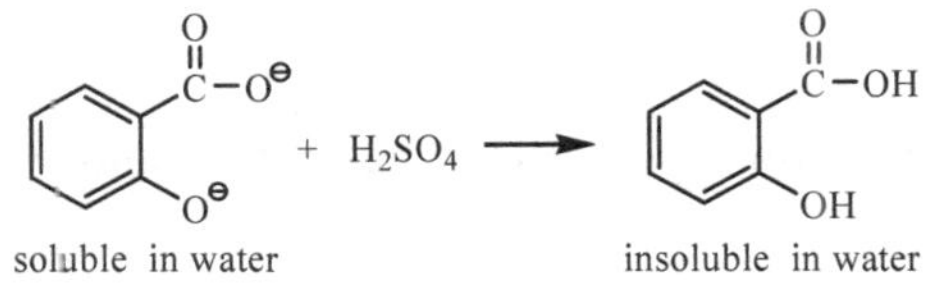

7. About 80.0 mL of water is enough for recrystallization.

8. The percent yield is 68~85 %.
The melting point of salicylic acid is 158.0~160.0℃.

9. Ester hydrolysis can be carried out by acid catalyst. It is a reversible reaction. An acid catalyzes a hydrolysis reaction by forming an oxonium ion on the oxygen atom of the ester group, which is susceptible to be attacked by the nucleophile.
It is the reverse reaction of an acid catalyzed esterification.

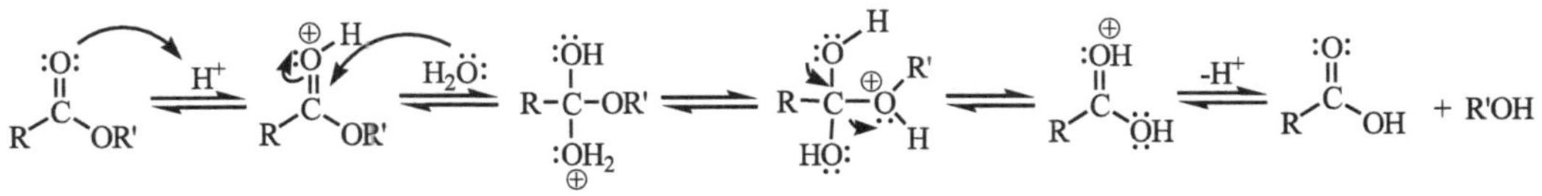

10. **Saponification Equivalent** is obtained by the weight of ester divided by moles of base used for saponification.

$$\text{Saponification Equivalent} = \frac{\text{weight of ester (in grams)}}{\text{mole of KOH}}$$

After heating a weighed sample of ester with an excess, yet known amount of potassium hydroxide solution, the reaction mixture is titrated with standardized hydrochloric acid to find the mole of the base used for saponification. It is the molecular weight of ester divided by the number of ester groups in the molecule.

Keywords for Phase Transfer Catalysts

Chemical equation for Synthesis of Salicylic Acid

Reaction mechanism for Synthesis of Salicylic Acid

2.3 Synthesis of Acetanilide

Keywords : protein, peptide bond, DMF, protecting group, solid phase synthesis

Proteins are large biomolecules chiefly held together by **amide** groups known as the **peptide linkages**. Thousands of enzyme required to catalyze biochemical reactions in living organisms belong to proteins. Proteins with fewer than 50 amino acids are often called **peptides**. About twenty different amino acids serve as the building blocks for **proteins**. They contain both basic amino groups (–NH_2) and acidic carboxyl groups (–COOH).

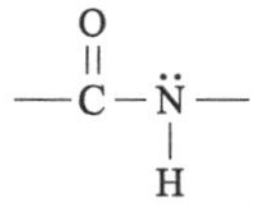

Peptide bond

The lone–pair electrons on nitrogen atom in an **amide** are not basic and the C–N bond behaves like a double bond because of the large resonance contribution. The chemical shifts of the methyl groups in DMF are different since the rotation of nitrogen–carbon bond is restricted.

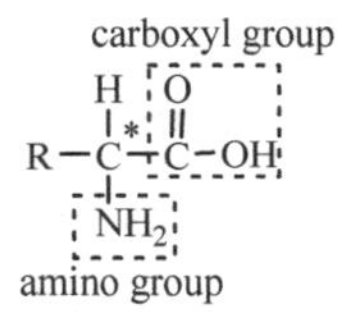

Structure of amino acid

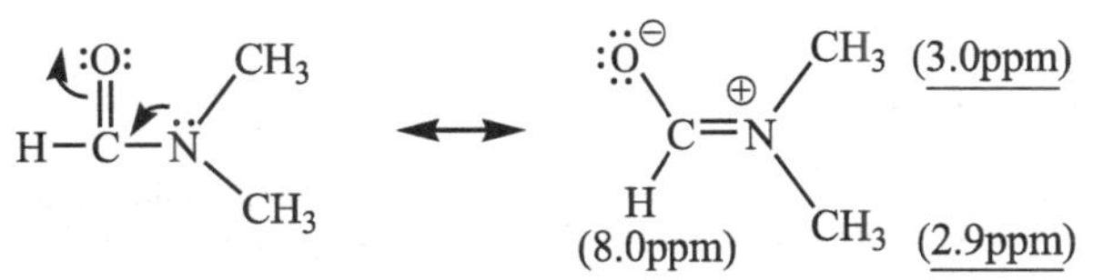

DMF (N,N -dimethyformamide)

Amides can be manufactured industrially by heating up the ammonium salt which is formed from carboxylic acid and amine at a high temperature

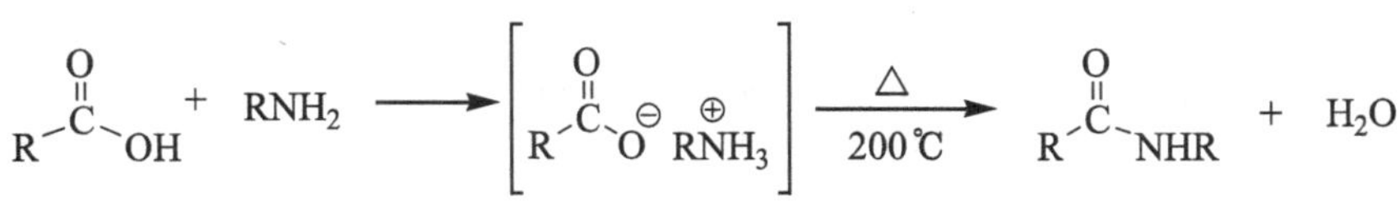

Since proteins need to bond in a certain order, the synthesis of a peptide bond can be carried out in the desired way by protecting the amino and carboxyl groups that are not necessary for the reaction. *tert*–Butoxy carbonyl (t–BOC) and benzyl carbamate (Cbz) are used for the **protection** of amino groups.

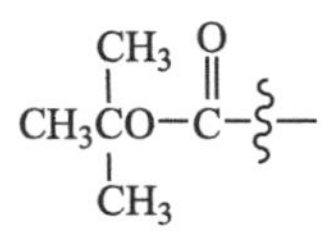

t - butoxy carbonyl

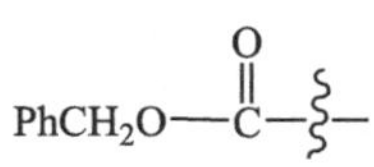

benzyl carbamate

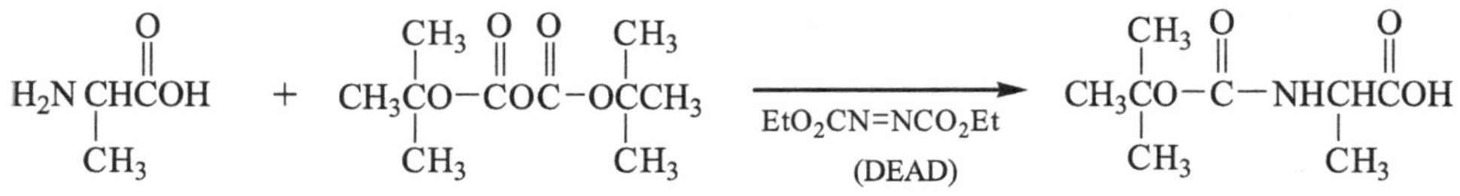

Alanine Di-tert-butyl dicarbonate t-BOC-Ala

A huge peptide chain can be synthesized by sequential additions of one amino acid at a time, using the solid phase method devised by **Merrifield.** The solid polymer protects the carboxyl group in the solid phase method. And it produces a high yield at each stage and has now been automated by the commercial peptide synthesizer.

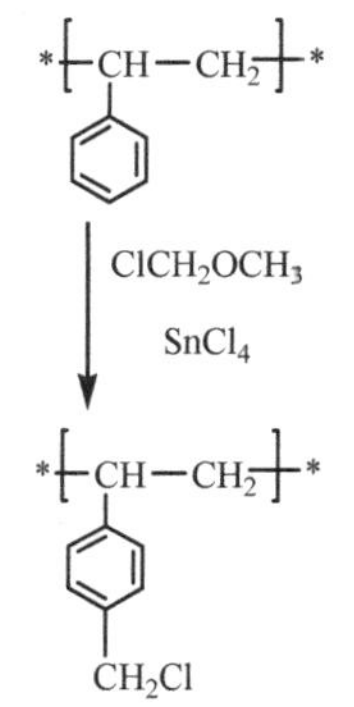

chloromethylation

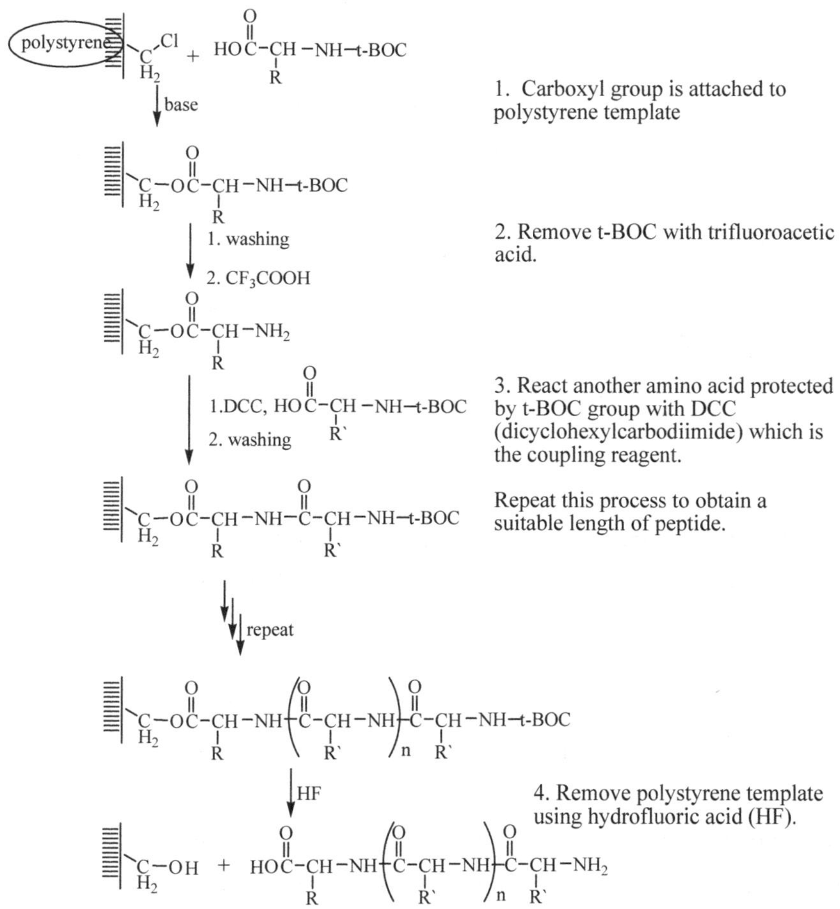

In this experiment, acetanilide is synthesized from aniline and acetic anhydride.

Chemical equation

Experiment

Reagents

Name	Molecular formula	Structural formula	M. W.	m.p. (℃)	b.p. (℃)	Density (g/mL)
Aniline						
Sodium acetate						
Acetic anhydride						
Acetanilide (Product)						

Apparatus

⇨ magnetic stirrer, magnetic spin–bar, lab jack, melting point apparatus, ice bath, UV lamp, vacuum pump, thermometer

Glassware

⇨ TLC chamber, 250 mL Erlenmeyer flask, 100 mL graduated cylinder, 50 mL beaker, 250 mL beaker, pipette, filter flask, Büchner funnel, glass rod

Experimental procedure

<Reaction>

1. Place 60.0 mL of water in a 250 mL Erlenmeyer flask **A** in an ice bath. Add 2.3 mL (27.5 mmol, 1 eq) of concentrated hydrochloric acid and warm it up to room temperature.
2. After 10 minutes, add 2.5 mL (2.60 g, 27.4 mmol, 1 eq) of aniline.
   To monitor the reaction, prepare 3 or 4 TLC plates and spot them with the starting material.
   Raise the temperature slowly up to 65~70℃. [Note 1]
   (Observation 1: change in solution, temperature)
3. Place 12.0 mL of water in a 50 mL beaker B and dissolve 2.30 g (28.0 mmol) of sodium acetate.
4. Add the solution in beaker **B** to the solution in flask **A**. Continue stirring while keeping the temperature at 65~70℃. [Note 2]
   (Observation 2: change in solution)
5. Add 6.0 mL (6.50 g, 63.6 mmol) of acetic anhydride. [Note 3]
6. After 20 minutes, check the reaction mixture by the pre–spotted TLC plate. <the developing solvent is ethyl acetate: hexane = 1:2>
   (Observation 3: check if aniline has disappeared on the TLC plate)
   When aniline is completely disappeared on TLC, continue stirring for another 10 minutes. Remove the heat source and allow the reaction flask to cool down to room temperature.
   (Observation 4: precipitation)
   Place the reaction flask in an ice bath.

Date of experiment : **Room temperature :**

Observation

Observation 1 :

Observation 2 :

Observation 3 :

Observation 4 :

<Work–up>

7. When no more crystal is precipitated, filter it by the appropriate suction filtration procedure that is dealt in **1.5 Recrystallization.** [Note 4]
   While filtering, prepare 100.0 mL of boiling water for recrystallization.
8. Keep a small amount of the crude product for comparison with the recrystallized product later on.
   Place the crude product in a 250 mL Erlenmeyer flask with the magnetic spin bar, and add boiling water in small portions until it dissolves completely. [Note 4] (Observation 5: amount of water, state of crystal before and after recrystallization)
9. When the solid dissolves completely, cool it down to room temperature, and then place it in an ice bath. The final product is obtained by the same suction filtration procedure mentioned above in the **procedure** 7.
10. After the product has sufficiently dried, weigh the product and calculate percent yield. Observe the melting point including the range and TLC of the crude product and the recrystallized product. [Note 5,6,7,8] (Observation 6: percent yield and melting point)

Observation 5 :

Observation 6 :

- Theoretical weight of product ____________ g
- Practical weight of product ____________ g
- Percent yield ____________ %

Checked by ______________________

Note

1. On adding aniline in the acidic aqueous solution, the oil drops float on the surface in the beginning but soon dissolve completely by the formation of a quaternary ammonium salt. The color of solution is pale yellow.

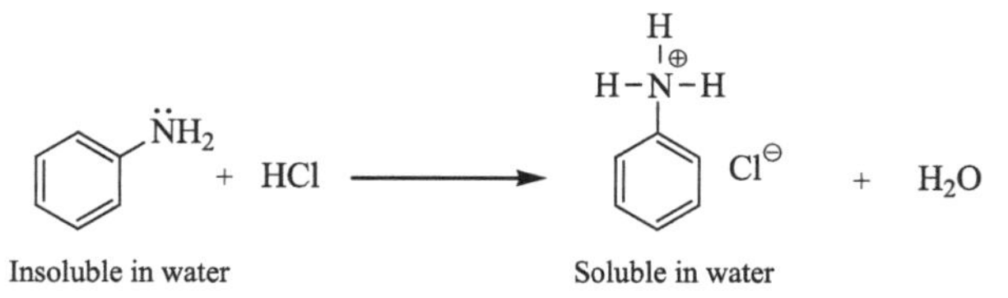

2. On adding sodium acetate, the color of solution becomes lighter. If sodium acetate is not added, crystallization will not be completed and the solution will stay at dark brown.

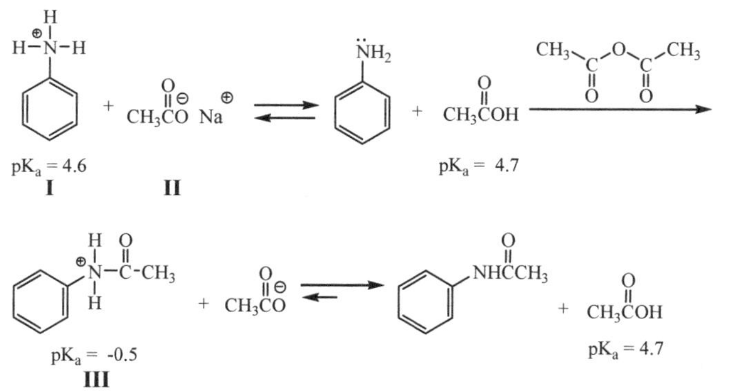

When hydrochloric acid is added to aniline, quaternary amine is formed which is soluble in water. The nucleophilicity of quaternary amine group drops significantly since it loses its lone–pair electrons. Sodium acetate is added to recover the nucleophilicity of amine in aniline. Since the acidities –pK_a– of both the acetic acid and the quaternary ammonium salt **I** of aniline are similar, aniline and acetic acid are formed in equilibrium with the quaternary salt and the carboxylate anion **II**.
The resulting aniline attacks acetic anhydride to produce the amide **III** salt and the carboxylate anion. Due to the large difference in pK_a, the resulting carboxylate anion readily abstracts the proton from the amide salt to give rise to acetanilide and acetic acid. The formation of a weak acid is the **driving force** for this reaction. (refer to **1.4 Extraction**)

3. On adding acetic anhydride, dark yellow oil drops will be formed but disappear shortly. If the temperature is not high enough, there might be a white solid and bubble in the reaction mixture. Add extra 1.0 mL of acetic anhydride if there is any starting material remained on TLC.

4. As the reaction proceeds, white crystals will be formed like layers of piled up thread. During recrystallization with hot water, a light brown colored impurity may be observed. It lowers the purity of the product.

5. The melting point of the white crystal obtained through recrystallization is 111.0~113.0℃. (lit. value : 113.0~115.0℃)
   The percent yield is 70~85 %.

6. It is interesting to note that the boiling points of the amides are much higher than those of the corresponding acids. In comparison, amines have lower boiling points than the corresponding alcohols.

   propionic acid: b.p. 141℃ | propionamide : b.p. 213℃
   butyl alcohol: b.p. 116~118℃ | butylamine : b.p. 78℃

7. Treating a primary amide with thionyl chloride gives rise to a nitrile compound.

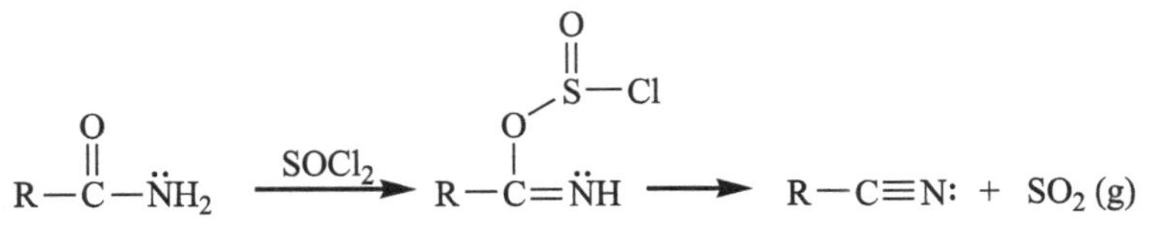

8. **Hofmann rearrangement**
   An amide is converted to a primary amine through isocyanate and carbamic acid intermediates.

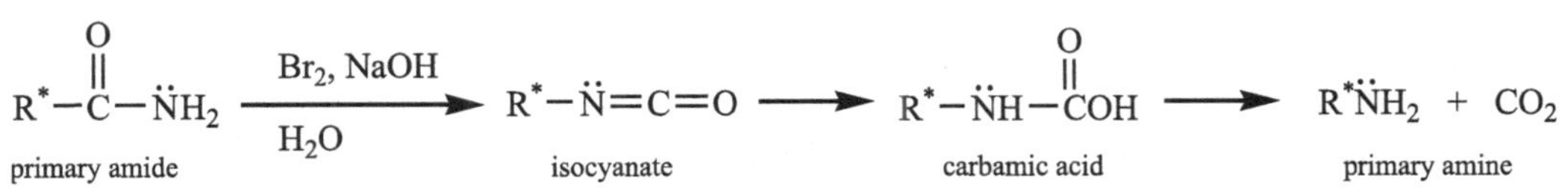

Keywords for Amides

Chemical equation for Synthesis of Acetanilide

Reaction mechanism for Synthesis of Acetanilide

2.4 Oxidation

Keywords : antioxidant, vitamin C, vitamin E, chromium reagent, Baeyer–Villger, Sharpless, Swern, filter cake

Typical examples of oxidation include the **combustion** reaction and the **metabolism** in organisms. The combustion of substance is a rapid oxidation reaction between oxygen and fuel, which produces heat and/or light. In an organism, the metabolism between the food taken and the oxygen from respiration produces energy, carbon dioxide, and water. The carbohydrate produces 4.0 kcal per 1.0 g through the metabolism and provides energy that we need for living.

Combustion: $(C, H)_n + O_2 \longrightarrow CO_2 + H_2O$ + Energy (heat or light)

In a living body : $C_n(H_2O)_n + O_2 \longrightarrow CO_2 + H_2O$ + Energy (heat)

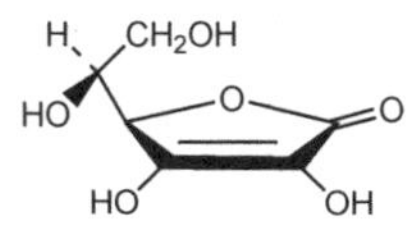

Ascorbic acid (vitamin C)

As we grow old, a long contact with oxygen causes oxidation such as the formation of wrinkles. An **antioxidant** is added to food or cosmetics to slow down the oxidation process. The well-known antioxidants in daily life include **vitamin C** (ascorbic acid) and **vitamin E** (tocopherol).

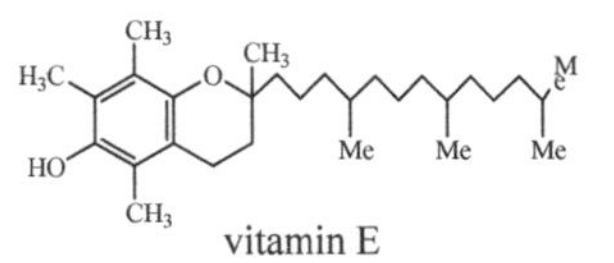

vitamin E

1. Oxidation using Chromium reagents

The most widely used transition metal oxidant is Cr (VI), which is made from chromium trioxide (CrO_3) or dichromate ($Cr_2O_7^{2-}$). Jones reagent (CrO_3 in H_2SO_4 or acetic acid with acetone) converts alcohol to carboxylic acid under the strong acidic condition. Oxidation of primary alcohols with a chromic acid is less satisfactory since it often produces ester as a by-product. Collins' reagent (CrO_3–2py), PCC (pyridinium chlorochromate), and PDC (pyridinium dichromate) provide milder reaction conditions compared to Jones reagent.

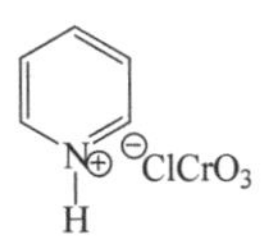

PCC

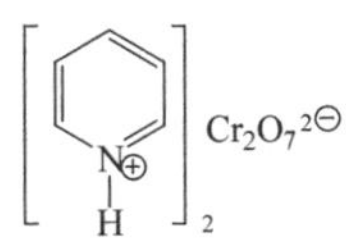

PDC

2. Baeyer–Villiger oxidation

This reaction takes place when an oxygen atom is added to a C–C bond to give lactone from ketone. In this reaction, the migration tendency is in the order of 3° alkyl> 2° alkyl > benzyl > phenyl > 1° alkyl.

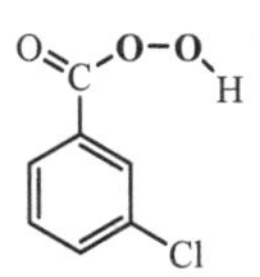

m-chloroperbenzoic acid (MCPBA)

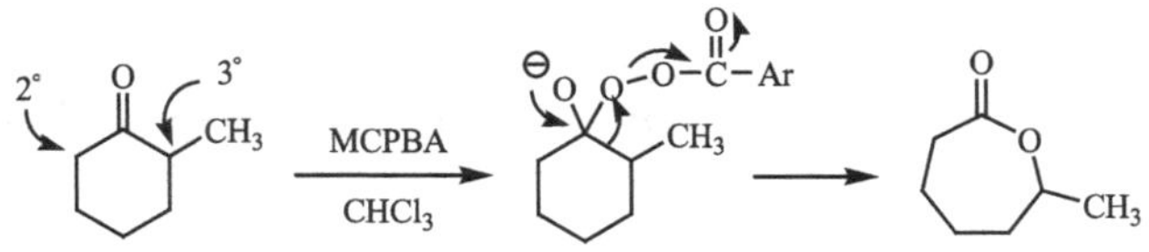

3. Sharpless epoxidation

The allylic alcohols yield optically active epoxy compounds in Sharpless epoxidation. t–Butyl hydroperoxide (t–BuOOH) provides oxygen atoms to the double bond. The chiral tartrate forms a coordinate covalent bond with a transition metal to provide a **chiral environment**.

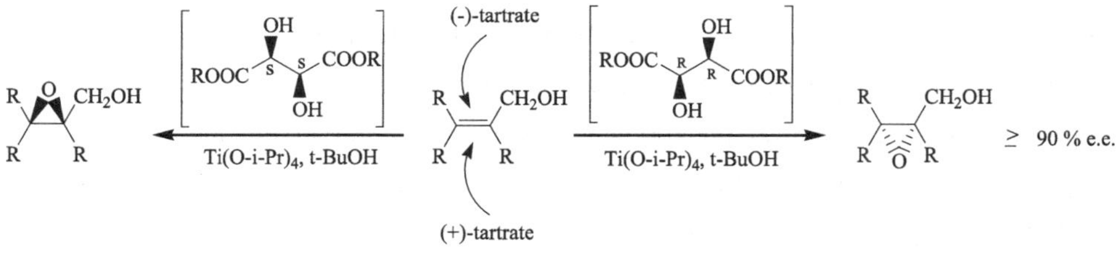

The hydroxy group in the allylic position bonds with the chiral coordinated complex above, producing more than 90 % of **enantiomeric excess (e.e.)**. It means producing 95 % and 5 % of enantiomers respectively. (95–5 = 90 % e.e)

4. Swern oxidation

Cr (VI), which has been used as an oxidant, is now known as one of **carcinogens**. Therefore it needs to be handled with care, and many oxidants have been developed to replace the Cr (VI). One of them is Swern oxidation, which is carried out to obtain aldehydes from 1° alcohols.

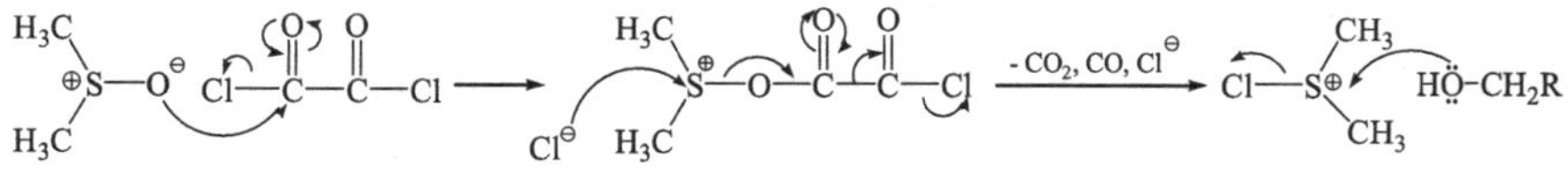

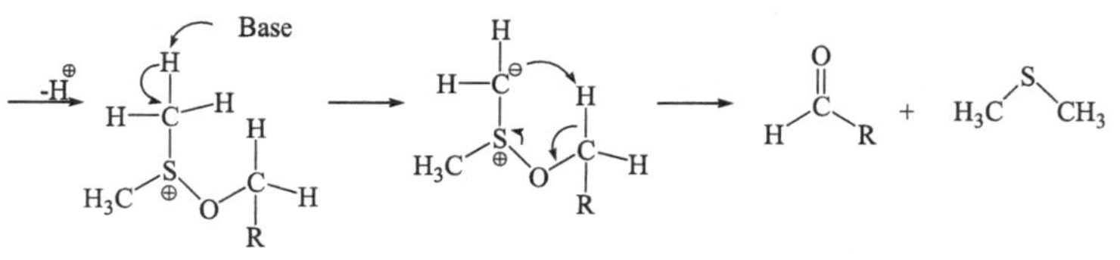

If there are bulky groups (R) bonded to the carbon atom where the hydrogen atom is intended for the **intramolecular deprotonation**, decomposition may happen and give rise to thio–ether as a by–product.

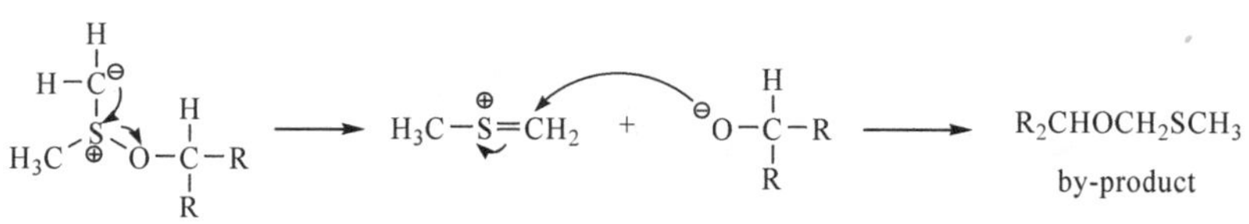

Chemical equation

Experiment I

Reagents

Name	Molecular formula	Structural formula	M. W.	m.p. (℃)	b.p. (℃)	Density (g/mL)
Cyclohexanol						
Sodium dichromate dihydrate						
Cyclohexanone (Product)						

Apparatus

⇨ magnetic stirrer, magnetic spin–bar, lab jack, UV lamp, ice bath, vacuum pump, evaporator, thermometer

Glassware

⇨ TLC chamber, 50 mL Erlenmeyer flask, 250 mL beaker, 100 mL graduated cylinder
250 mL separatory funnel, 100 mL Erlenmeyer flask, 100 mL one–neck round bottom flask, 500 mL one–neck round bottom flask, filter flask, Büchner funnel

Experimental procedure

<Reaction>

1. Place 15.8 g (53.0 mmol) of sodium dichromate dehydrate ($Na_2Cr_2O_7 \cdot 2H_2O$) in a 50 mL Erlenmeyer flask **A** and dissolve in 15.0 mL water. [Note 1, 2]

2. Place 25.0 g of ice slurry and the spin–bar in a 250 mL beaker **B** with the thermometer. Slowly add 10.0 mL (18.4 g, 187.6 mmol, 5.31 eq) of concentrated sulfuric acid and stir well.

3. Add 20.0 mL hexane and 10.0 mL cyclohexanol (9.48g, 94.6 mmol, 1.00 eq) to the sulfuric acid solution in beaker **B** and stir thoroughly. [Note 3]
   To monitor the reaction, prepare 3 or 4 TLC plates and spot them with the starting material.

4. Add 10.0 mL (35.3 mmol, 1.10 eq) of the prepared chromate solution in flask **A** to the beaker **B** in an ice bath. As the temperature rises during the addition, adjust the amount of the chromate solution in order to keep the temperature of the reaction mixture at around 45~50℃. [Note 4,5]
   (Observation 1: observe the temperature change of the solution)

5. When cyclohexanol completely disappears on TLC, continue stirring for another 10 minutes. <the developing solvent is ethyl acetate: hexane = 1:4> [Note 6] While keeping the beaker **B** in an ice bath, add 30.0 mL of diethyl ether to the reaction mixture, and then add 1.00 g (11.1 mmol) of oxalic acid. [Note 7,8]
   (Observation 2: TLC)

6. Remove beaker **B** from the ice bath and allow it to warm up to room temperature with stirring. After stirring for 10 minutes at room temperature, remove the spin–bar and allow it to stand undisturbed until the layers are completely separated.

Date of experiment : **Room temperature :**

Observation :

Observation 1 :

Observation 2 :

<Work–up>

7. Transfer the reaction mixture to the 250 mL separatory funnel **C** and collect the aqueous layer to another 250 mL separatory funnel **D**. If cyclohexanone remains in the aqueous layer on checking TLC, the aqueous layer should be extracted with 30.0 mL of diethyl ether. <the developing solvent is ethyl acetate: hexane = 1:4> (Observation 3: TLC)

8. The organic layers are combined in the 250 mL separatory funnel **C**. If the color of the organic layer is green, wash it twice with 10.0 mL of water each time. It will turn transparent yellow.
Wash the organic layer with 30.0 mL of saturated sodium chloride solution (brine) and separate the organic layer.

9. The organic layer is poured through the mouth of the separatroy funnel into a 250 mL Erlenmeyer flask **E**. Rinse the empty se paratroy funnel with 10.0 mL diethyl ether.
Add a small amount of anhydrous magnesium sulfate ($MgSO_4$) to the organic layer in the Erlenmeyer flask **E** until the solution becomes clear and the solid $MgSO_4$ moves freely in the solution. The resulting organic layer is filtered according to the appropriate suction filtration procedure, which is dealt in **1.5 Recrystallization**.

10. Transfer the resulting solution (**filtrate**) in a small portion to a pre–weighed 100 mL one neck round bottom flask (**tared flask**). Evaporate diethyl ether by using a rotary evaporator. When the solvent is almost evaporated, cyclohexanone will remain at the bottom of the flask.

11. Check the purity of the product with TLC, and calculate the percent yield. [Note 9, 10,11,12,13,14] (Observation 4: TLC)

Observation 3 :

Observation 4 :

- Theoretical weight of product ____________ g
- Practical weight of product ____________ g
- Percent yield ____________ %

Checked by ______________________

Note

1. The oxidizing chromic solution contains a variety of species in equilibrium. The most important Cr (VI) species is the acid **chromate ion**, $HCrO_4^-$. Cr (VI) is ultimately reduced to Cr (III) by several steps.

$$\underset{\text{dichromate}}{H_2O + Cr_2O_7^{2-}} \rightleftharpoons \underset{\text{acid chromate}}{2HCrO_4^-}$$

2. When handling chromium reagents, wear gloves and avoid any contact with skin. In case of contact with skin, wash immediately and sufficiently with running water.

3. An addition of hexane can prevent the reaction mixture from an abrupt heat rise and make the monitoring of TLC easier.

4. The oxidation by dichromate is an exothermic reaction; therefore dichromate should be slowly added to an ice bath.

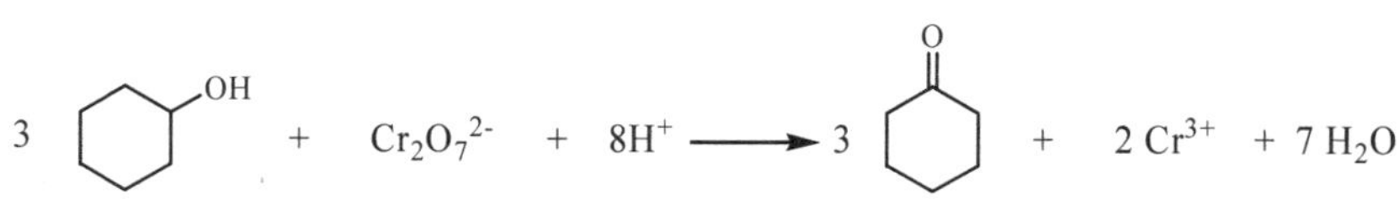

5. Although 10.0 mL is the required, a little extra amount will not harm the reaction. Therefore, add enough dichromate until the reaction is complete by checking with TLC. It cannot be decided whether the reaction is complete just by observing the color of the chromium ion. The change in color only gives us the qualitative information –not the quantitative one– so it is essential to check by TLC.

6. As the cyclohexanol and the product are not visible under the UV lamp, use potassium permanganate ($KMnO_4$) dyeing reagent to check the formation of the product.

7. The oxalic acid breaks down an excess amount of dichromate in the solution.

$$Cr_2O_7^{2-} + 3\,HO{-}\overset{\overset{\displaystyle O}{\|}}{C}{-}\overset{\overset{\displaystyle O}{\|}}{C}{-}OH + 8H^+ \longrightarrow 2Cr^{3+} + 6CO_2 + 7H_2O$$

8. Diethyl ether should be added at the end of the reaction to help the separation of the product from the chromate salt since the solubility of chromate salt is low in diethyl ether.

9. If you check TLC after working up the reaction or after concentrating the product, there may be little spots appeared under the spot of cyclohexanone. However, it is acceptable for our purpose of this experiment. <R_f of cyclohexanol = 0.4 and R_f of cyclohexanone = 0.66>

10. The percent yield is 65~75 %.
    The yield obtained by the evaporated product is higher since diethyl ether is not completely removed by a rotary evaporation.

11. The oxidation of alcohol by chromic acid is believed to take place by fast initial formation of **chromate ester** followed by slow deprotonation of carbon, which is attached to the hydroxyl group. In consequence, the resulting chromate ester produces aldehydes from primary alcohols, and ketones from secondary alcohols. Tertiary alcohols do not undergo oxidation since they do not have a proton to be abstracted.

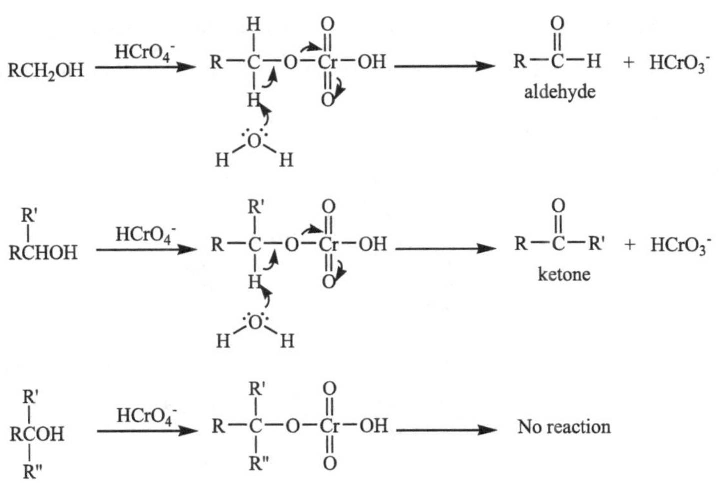

12. The aldehyde and the hydrated aldehyde are further attacked by the acid chromate to produce a carboxylic acid.

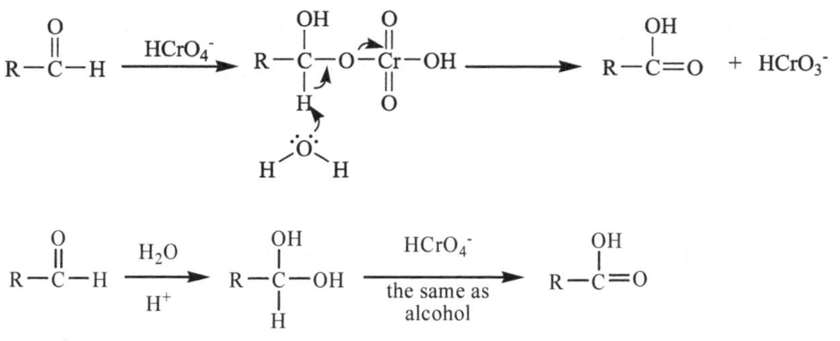

13. The control of oxidation of primary alcohols to aldehydes under the acidic solutions of chromic acid is usually less satisfactory. The resulting aldehyde is easily oxidized further to carboxylic acid (**Note 12**) and/or reacts with unchanged alcohol to form a **hemiacetal** which is readily oxidized to an ester.

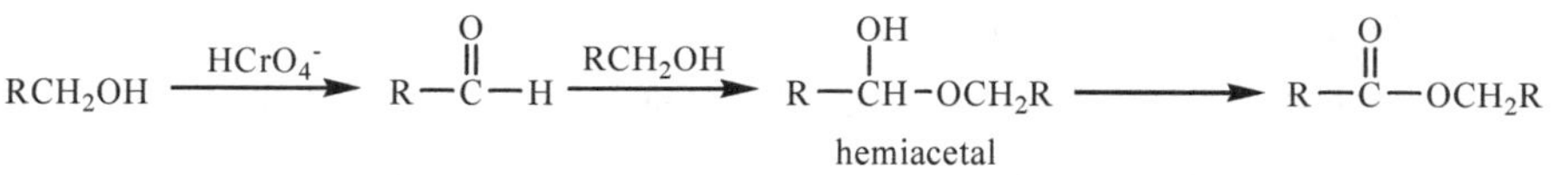

14. For the alcohols attached to the cyclohexane skeleton, it has been found that an **axial hydroxyl group** is generally oxidized more rapidly than an **equatorial hydroxyl group** by a factor of about 3. Since **1,3-diaxial steric strain** in an axial chromate ester is substantially relieved by going from the reactant to the product, an axial chromate ester is decomposed more rapidly than an equatorial chromic ester.

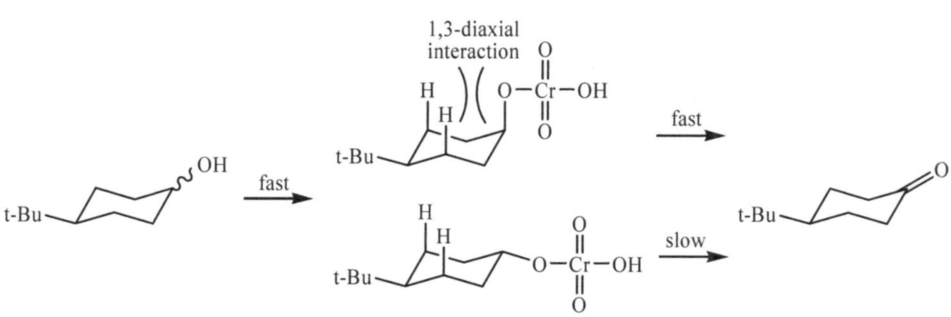

Chemical equation

Experiment 2

Reagents

Name	Molecular formula	Structural formula	M. W.	m.p. (℃)	b.p. (℃)	Density (g/mL)
Pyridinium chlorochromate (PCC)						
Benzyl alcohol						
Benzaldehyde (Product)						

Apparatus

⇨ magnetic stirrer, magnetic spin–bar, lab jack, UV lamp, vacuum pump, evaporator

Glassware

⇨ TLC chamber, 250 mL one–neck round bottom flask, 500 mL one–neck round bottom flask, 100 mL graduated cylinder
500 mL separatory funnel, 100 mL beaker, pH paper, 500 mL filter flask, Büchner funnel

Experimental procedure

<Reaction>

1. Place 6.24 g (29.0 mmol, 1.5 eq) of PCC and 70 mL of methylene chloride (M.C.) in a 250 mL one–neck round bottom flask **A** and stir. Check pH of the solution using pH paper. [Note 1] (Observation 1: state when PCC is added to methylene chloride, color of pH paper)

2. Add 2.0 mL (2.10 g, 19.3 mmol, 1.0 eq) of benzyl alcohol to 20.0 mL of methylene chloride in flask **A** and stir thoroughly. To monitor the reaction, prepare 3 or 4 TLC plates and spot them immediately after the addition. [Note 2] <the developing solvent is ethyl acetate : hexane = 1:2> (Observation 2 : color of the solution on adding benzyl alcohol and TLC)

3. Monitor the reaction with TLC every 20 minutes. [Note 3] (Observation 3: TLC)

4. When benzyl alcohol completely disappears on TLC, continue stirring for another 10 minutes. Add 25.0 mL of diethyl ether twice, and stir for another 10 minutes. [Note 4]

<Work–up>

5. Add 7.00 g of Celite and 25.0 mL of diethyl ether with stirring. [Note 5] (Observation 4: state before and after adding Celite.)

6. Prepare a filter cake to filter the reaction mixture.
   Spread anhydrous magnesium sulfate ($MgSO_4$) about 1 cm thick on the filter paper in Büchner funnel. Soak with diethyl ether under the aspirator suction and level it. Place a filter paper on the resulting magnesium sulfate and add silica gel dissolved in the diethyl ether until it is 2 cm thick. Level it even under the aspirator suction. Lastly, put another filter paper on the silica gel and filter the reaction mixture according to the appropriate suction filtration procedure, which is dealt in **1.5 Recrystallization**. [Note 6] (Observation 5: color of filtrate)

Date of experiment : **Room temperature :**

Observation

Observation 1 :

Observation 2 :

Observation 3 :

Observation 4 :

Observation 5 :

7. Wash the filter cake with 50.0 mL of diethyl ether. [Note 7] Transfer the filtrate to a 500 mL one–neck round bottom flask **B**.
8. Add 50.0 mL of 2.5 M sodium hydroxide solution to the filtrate in the flask **B** and stir well. Transfer the resulting mixture to a 500 mL separatory funnel and separate the water layer. Add 50.0 mL of 2.5 M sodium hydroxide solution to the organic layer and extract once again. [Note 8] (Observation 6: examine whether there is separation between the aqueous and the organic layers and compare their colors)
9. Wash the organic layer with 50.0 mL of distilled water followed by 30.0 mL of brine. The organic layer is poured through the mouth of the separatory funnel into a 500 mL Erlenmeyer flask **C**. Rinse the empty separatory funnel with 10.0 mL of diethyl ether.
10. Add a small amount of anhydrous magnesium sulfate ($MgSO_4$) to the organic layer in the Erlenmeyer flask C until the solution becomes clear and the solid $MgSO_4$ moves freely in the solution. The resulting organic layer is filtered according to the appropriate suction filtration procedure, which is dealt in **1.5 Recrystallization**.
11. Transfer the resulting solution (**filtrate**) in a small portion to a pre–weighed 100 mL one neck round bottom flask (**tared flask**). Evaporate diethyl ether by using a rotary evaporator. When the solvent is almost evaporated, benzaldehyde will remain at the bottom of the flask
12. Check the purity of the product using TLC, and calculate the percent yield. [Note 9,10,11,12] (Observation 7: percent yield)

Observation 6 :

Observation 7 :

- Theoretical weight of product ____________ g
- Practical weight of product ____________ g
- Percent yield ____________ %

Checked by ______________________

Note 2

1. PCC floats in the methylene chloride solution in the powder form. On adding PCC reagent, the pH changes to the region of a weak acid.

2. Oxidation happens immediately on adding benzyl alcohol. As a result, chromate salt is produced and the color of solution turns dark brown. Since the reaction starts immediately, spot the TLC plate at once and check how long the reaction has progressed. The resulting salt can interfere with the stirring, so spin the spin–bar as quickly as possible.

3. It takes about 30 minutes until the product disappears completely. If the product is remained on the TLC plate, check the reaction mixture every 10 minutes. Stir well as the reaction progresses. Use a glass rod if the spin–bar does not spin well. A small amount of benzoic acid would be observed on the TLC along with the expected product (benzaldehyde). Benzoic acid can be removed by extraction with a basic solution in the work–up process.

4. A small amount of PCC dissolved in the solution can be precipiated in the solid form by adding diethyl ether which has no solubility against PCC.

5. **Celite** laid on the filter paper is one of the filter aids, which adsorb inorganic impurities. In this experiment, add the same or a little excess amount of Celite as PCC used in the reaction. It will help removing the chromate salts by adsorbing the reduced chromate salts on its surface.

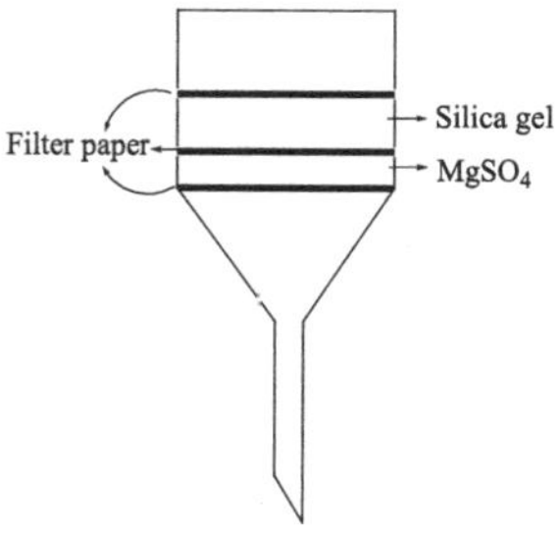

6. A small amount of salt in the solvent can be filtered with anhydrous $MgSO_4$ and silica gel in this process. Especially the silica gel strongly adsorbs the polar compounds, and therefore it is recommended that the thickness of silica gel should be thicker than that of magnesium sulfate.

7. Wash the product adsorbed on the surface of the salt with diethyl ether.

8. When the chromate salt is not completely removed by filtration, adding a basic solution to the organic layer can separate the salts which is dissolved in the basic aqueous layer. The by–product, benzoic acid, can also be removed by this process.

9. The percent yield is 65~75 %. If the amount of reactant is reduced, the percent yield will also decrease. The yield obtained by the evaporated product is higher since diethyl ether is not completely removed by a rotary evaporation. And if the work–up process is carried out carelessly, the percent yield will significantly drop to under 40 %.

10. PCC is made by dissolving chromium trioxide in hydrochloric acid and adding the basic solvent pyridine to the solution. However, the mildly acidic nature of PCC may preclude its use with acid–sensitive compounds.

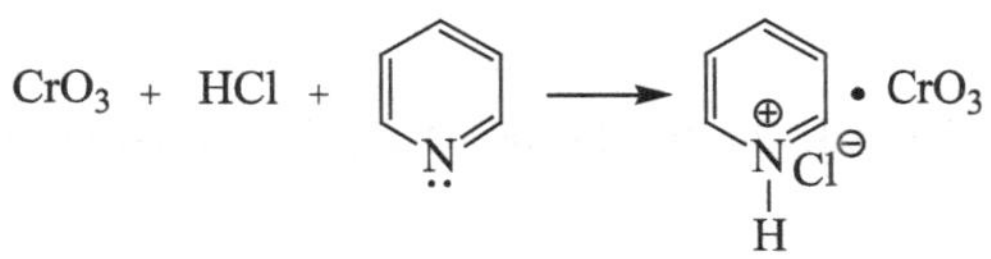

11. The PCC oxidation undergoes with CrO_3 related reaction intermediate under anhydrous condition. Under an **anhydrous condition**, hydrated aldehyde and acid chromate ion are not formed since water is not available. As a result, oxidation stops at the aldehyde stage. (refer to the previous section **Note 1 and 12**)

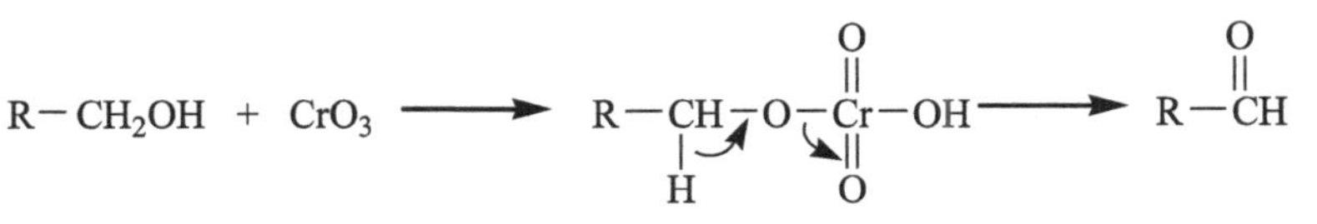

12. Another good reagent is pyridinium dichromate (PDC) in CH_2Cl_2 or DMF solution which oxidizes alcohols containing an acid sensitive group –e.g. double bond– to aldehydes or ketones, especially allylic alcohols to unsaturated carbonyl compounds.

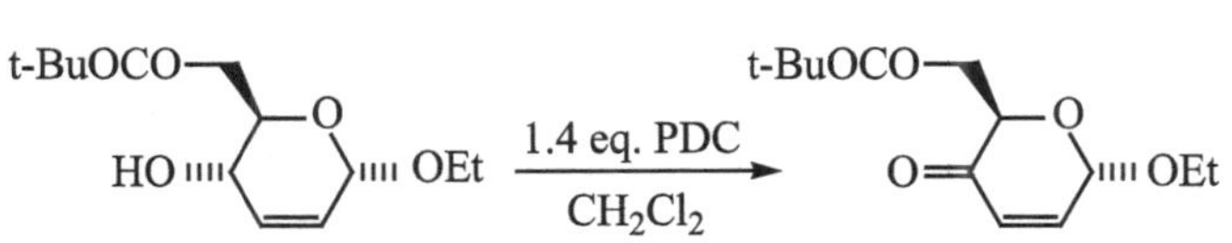

Keywords for Oxidation

Chemical equation for Synthesis of Cyclohexanone and Benzaldehyde

Reaction mechanism for Synthesis of Cyclohexanone and Benzaldehyde

2.5 Reduction

Keywords : Lindlar catalyst, poison, Birch reduction, *exo* / convex, *endo* / concave, stereoselectivity

During **photosynthesis**, green plants absorb energy from the sun to make glucose and oxygen from carbon dioxide and water. In the process of photosynthesis, the carbon in carbon dioxide is reduced to make sugar, and the oxygen in water is oxidized to produce oxygen gas.

$$6CO_2 + 24H^+ + 24e^- \longrightarrow C_6H_{12}O_6 + 6H_2O$$

$$12H_2O \longrightarrow 6O_2 + 24H^+ + 24e^-$$

$$6CO_2 + 6H_2O \xrightarrow{h\nu} C_6H_{12}O_6 + 6O_2$$

The oxidation and reduction reactions always take place simultaneously. The energy stored in glucose is transferred eventually to the molecules such as ATP, which releases energy to drive other chemical reactions.

The reduction in organic reaction can be classified by the reactive species.

1. Gain of hydrogen

The efficiency of catalytic hydrogenation is influenced by the presence of organic and/or inorganic complex. The addition of lead acetate to palladium on calcium carbonate – **Lindlar catalyst** – slows down the uptake of hydrogen on the triple bond to exclusively produce a *cis*–double bond. Sulfur and most sulfur containing compounds are poisons, which strongly inhibit hydrogenation.

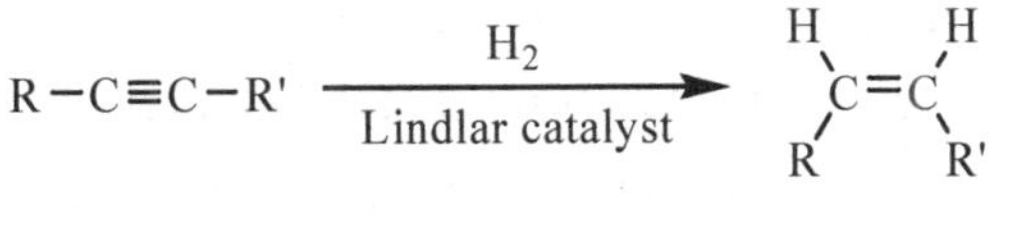

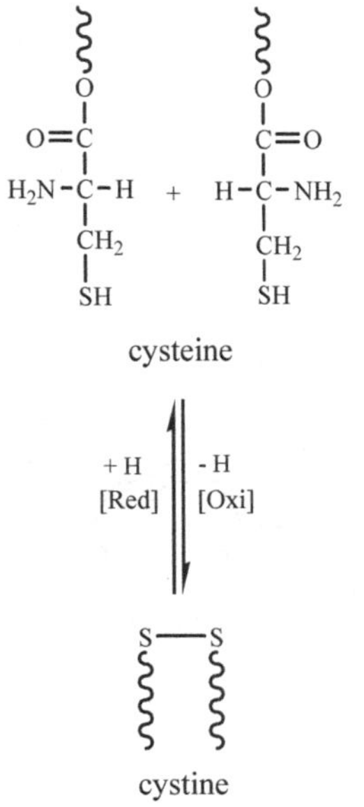

cysteine

cystine

Cystine plays a special role as the cross–linker in the protein structure since the covalent disulfide linkage joins the two **cysteine** half–residues. This linkage undergoes a cleavage by the action of reducing agents and induces a change in the protein structure. This transformation is reversible and important in biochemistry.

2. Gain of electron (Birch reduction)

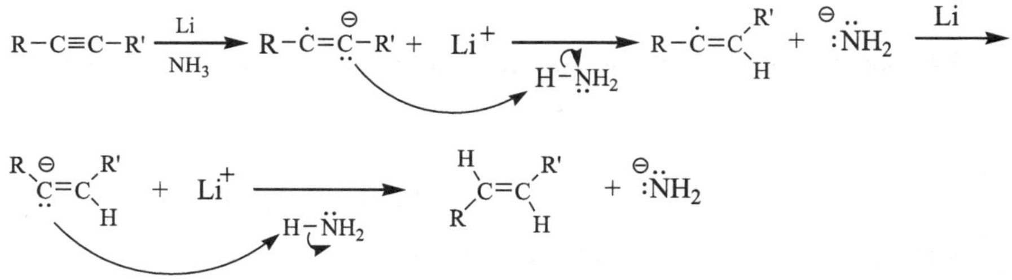

Because of the stepwise acceptance of electrons, the net result of reductions is usually *anti* or *trans* addition of hydrogen. Birch reduction is especially successful when the source of protons –ethanol, *t*–butyl alcohol, or compounds with acidic hydrogen– is present in the reaction medium. Its function is to protonate the intermediate anion radicals and thus to cut down any undesirable side reactions such as dimerization of the radical anion and the polymerization.

3. Loss of oxygen (electrolytic reduction)

Nitrobenzene gives rise to an azo compound by electrolytic reduction on the surface of Mg electrodes under the constant current. The important factors for electrolytic reduction are the nature of metals (Mg), electrolytes ($LiClO_4$), solvent (THF) and the amount of current passing through the electrolytic cells.

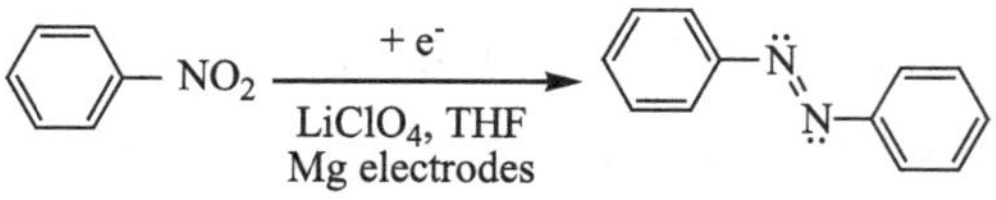

4. Reaction with hydride complex (H:⁻)

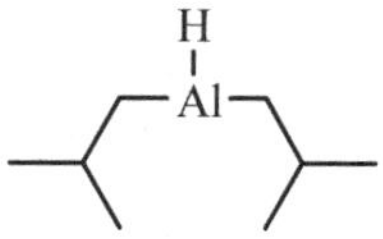

DIBAH

The typical regents used in the reduction of carbonyl groups are $NaBH_4$, $LiAlH_4$, and DIBAH (diisobutylaluminum hydride). $LiAlH_4$ reduces aldehyde, ketones, carboxylic acid, ester and amide to their corresponding alcohols. It reacts violently in alcohol and water to generate hydrogen gas. DIBAH reduces acid, ester and nitrile to aldehyde at a low temperature.

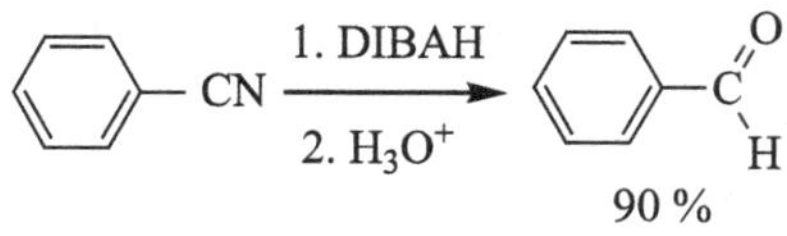

Chemical equation

Experiment I

Reagents

Name	Molecular formula	Structural formula	M. W.	m.p. (℃)	b.p. (℃)	Density (g/mL)
Benzaldehyde						
Sodium borohydride						
Benzyl alcohol (Product)						

Apparatus

⇨ magnetic stirrer, magnetic spin–bar, lab jack, UV lamp, ice bath, heating mentle, transformer, vacuum pump, evaporator

Glassware

⇨ TLC chamber, 100 mL two–neck round bottom flask, 100 mL graduated cylinder, reflux condenser
100 mL separatory funnel, 250 mL separatory funnel, 250 mL Erlenmeyer flask, 100 mL one–neck round bottom flask, filter flask, Büchner funnel

Experimental procedure

<Reaction>

1. Prepare a 100 mL two–neck round bottom flask **A** in an ice bath, and place 3.0 mL(3.10 g, 29.5 mmol, 1.0 eq) of benzaldehyde and 30.0 mL of methanol.
   To monitor the reaction, prepare 3 or 4 TLC plates and spot them with the starting material.
2. Add 700.0 mg (18.5 mmol, 0.6 eq) of $NaBH_4$ in small portions to the flask **A**. Stir for 15 minutes and check the TLC. <the developing solvent is ethyl acetate: hexane = 1:4> [Note 1, 2, 3]
   (Observation 1: temperature of the liquid, TLC)
3. When benzaldehyde has completely disappeared on the TLC, continue stirring for another 10 minutes.
   Add 10.0 mL of water, and then remove the ice bath.
   Stir for 10 minutes at room temperature. [Note 4]
4. Assemble the reflux apparatus on the flask **A**.
   Reflux for 10 minutes and cool down to room temperature.
   (Observation 2: pH)

<Work–up>

5. Add 30.0 mL of methylene chloride to the reaction mixture in the flask **A** and stir well.
6. Transfer the organic layer to 250 mL Erlenmeyer flask **B.** Check the aqueous layer with TLC, and if the product remains, extract by adding another 30.0 mL of methylene chloride. [Note 5]
   (Observation 3: TLC)
   Transfer the combined organic layer to the flask **B**.
7. Add 30.0 mL of saturated $NaHCO_3$ soution to the combined organic layer in the flask **B** and stir vigorously to remove acidic impurities.
   (Observation 4: gas evolution and layer separation)

Date of experiment : **Room temperature :**

Observation

Observation 1 :

Observation 2 :

Observation 3 :

Observation 4 :

8. Transfer it to the separatory funnel and shake well. And then extract the organic layer. The organic layer is drawn off carefully through the stopcock into another separatroy funnel. Rinse the stem with 10.0 mL of methylene chloride. The organic layer is washed with brine and then drawn off carefully through the stopcock into a clean Erlenmeyer flask **C**. Rinse the stem with 10.0 mL of methylene chloride. Add a small amount of anhydrous sodium sulfate (Na_2SO_4) to the organic layer in an Erlenmeyer flask **C** until the solution becomes clear and the solid Na_2SO_4 moves freely in the solution.
9. Filter the resulting organic layer using the appropriate suction filtration procedure, which is dealt in **1.5 Recrystallization**.
10. Transfer the resulting solution (**filtrate**) in a small portion to a pre–weighed 100 mL round bottom flask (**tared flask**). Evaporate methylene chloride by using a rotary evaporator. When the solvent is almost evaporated, benzyl alcohol will remain at the bottom of the flask.
11. Weigh the flask and check the purity of the product by TLC. Calculate the percent yield [Note 6~12]
(Observation 5: TLC, weigh and percent yield)

Observation 5 :

- Theoretical weight of product ____________ g
- Practical weight of product ____________ g
- Percent yield ____________ %

Checked by ______________________

Note I

1. Sodium borohydride readily absorbs moisture in the air because of its deliquescent property. Weigh it in the beaker or in the watch glass, and not on a weighing paper.
   Make sure to add it to the reaction flask not too slowly.
   When adding sodium borohydride, place the reaction flask in an ice bath since it is an exothermic reaction.

2. The four hydrogen atoms of sodium borohydride can be used as a hydride. In theory, 1 mol of sodium borohydride can reduce 4 mol of aldehyde. However, considering the loss of $NaBH_4$ reacting with the solvent and the moisture in the air, use an excess amount of sodium borohydride. (0.5 eq instead of 0.25 eq.)

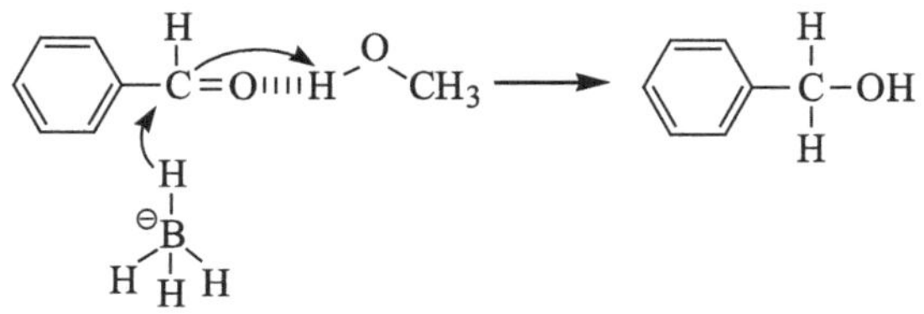

3. If benzaldehyde is remained on checking TLC, add another 570 mg (0.5 eq) of sodium borohydride.

4. When water is added, the compound of tetraalkoxyboron produces methyl alcohol and $B(OH)_3$. The reaction mixture is basic.

$$\left(H_3CO\right)_4 B^{\ominus}\,Na^{\oplus} + 4H_2O \xrightarrow[\Delta]{} BO_3^{3-} + 4\,CH_3OH + Na^{\oplus}$$

5. When using methylene chloride, be aware that its density is heavier than water, so the organic layer is placed below the aqueous layer. If a small amount of methylene chloride is used, the layers may not be divided well. After extracting twice with 30.0 mL of methylene chloride, use TLC to check no product is left in the aqueous solution. If the product is remained, extract once more with 20.0 mL of methylene chloride.

6. 80~90 % of the yield is to be obtained. The yield obtained by the evaporated product is higher since methylene chloride is not completely removed by a rotary evaporation.

7. The direction of hydride attack on the carbonyl group in **[2.2.1] bicyclic system** is influenced by the group on the bridge carbon. If hydrogen is attached in the bridge carbon, hydride attacks from the ***exo* face** or **convex** to produde *endo* alcohol. The *exo* face is less hindered than the *endo* face.
If the methyl group is attached in the bridge carbon, hydride attacks from the ***endo* face** or **concave** to produce the *exo* alcohol. The *exo* face is blocked by the methyl group on the bridge carbon.

I, II

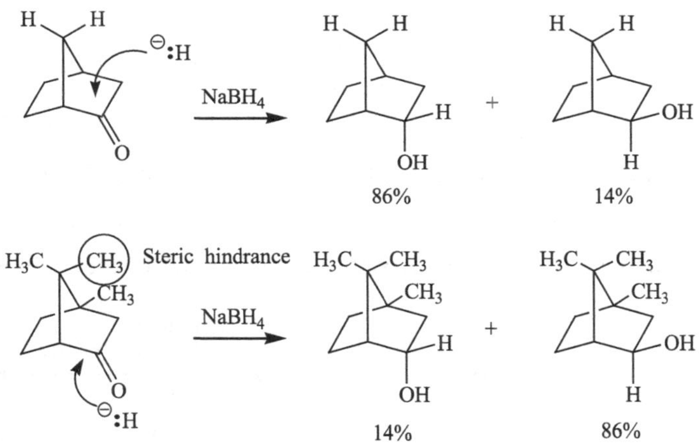

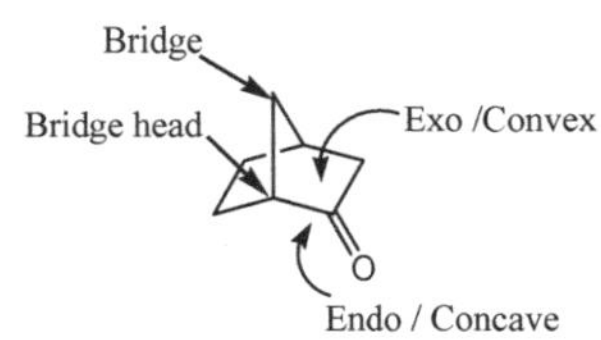

8. Sodium borohydride is less reactive than lithium aluminium hydride. As a result, it is more discriminating in its action. (**chemoselectivity**) It readily reduces aldehydes and ketones at room temperature but it does not generally attack esters or amides. Therefore it is usually possible to reduce aldehydes and ketones selectively in the presence of a variety of other functional groups.
It usually undergoes reduction in an alcoholic solutions since it reacts slowly with alcohols.

9. $NaBD_4$ (D:deuterium) and $NaBT_4$ (T:tritium) are available for labeling.

10. In the presence of cerium (III) chloride, sodium borohydride reduces *α*, *β*–unsaturated ketones to give allylic alcohols with a high selectivity (1,2–reduction > 1,4–reduction).

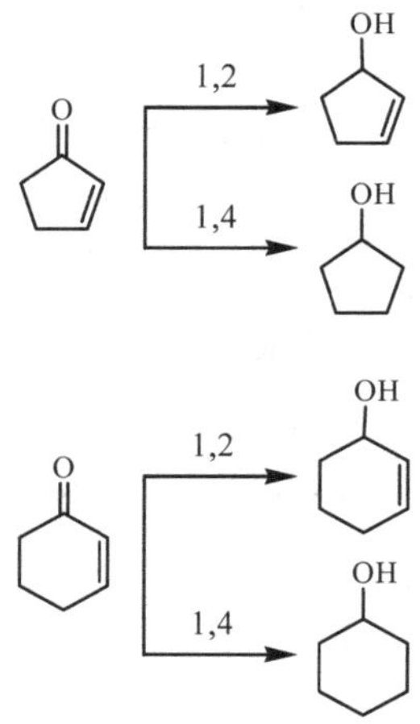

	$NaBH_4$	$NaBH_4/CeCl_3$
1,2	0 %	97 %
1,4	100 %	3 %
1,2	59 %	> 99 %
1,4	41 %	1 %

11. Sodium cyanoborohydride is less reactive than $NaBH_4$ and stable in aqueous solutions at pH > 3. It is used in the reductive amination of **iminium salts** to produce secondary amines.

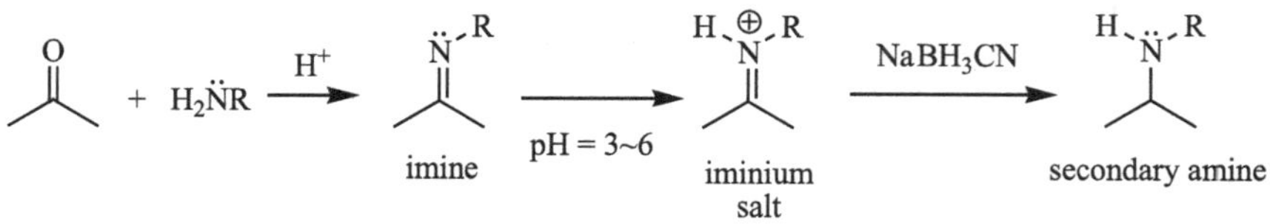

12. Lithium borohydride is more reactive than $NaBH_4$. In general, lithium compounds are more reactive than the corresponding sodium compounds since Li^+ coordinates to oxygen stronger than Na^+.

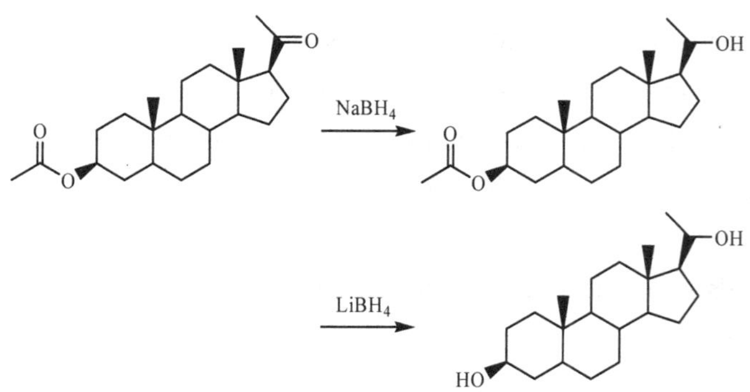

Chemical equation

Experiment 2

Reagents

Name	Molecular formula	Structural formula	M. W.	m.p. (℃)	b.p. (℃)	Density (g/mL)
Nitrobenzene						
Tin (Sn)						
Aniline (Product)						

Apparatus

⇨ magnetic stirrer, magnetic spin–bar, lab jack, UV lamp, ice bath, heating mentle, transformer, vacuum pump, evaporator

Glassware

⇨ TLC chamber, 100 mL two–neck round bottom flask, 3–way coke, 100 mL graduated cylinder, reflux condenser
distillation set, 100 mL separatory funnel, 100 mL Erlenmeyer flask, 100 mL one–neck round bottom flask, filter flask, Büchner funnel

Experimental procedure

– All procedure should be done under the hood –

<Reaction>

1. In a 100 mL two–neck round bottom flask **A**, place 5.0 mL (6.00 g, 48.6 mmol, 1.0 eq) of nitrobenzene and 12.0 g (101. mmol, 2.1 eq) of tin (Sn). Heat the reaction mixture by boiling water bath with stirring. [Note 1]
2. Add 15.0 mL (18.0 g, 172.3 mmol, 3.6 eq) of concentrated hydrochloric acid slowly in 2 mL portions. As the reaction proceeds, add 15.0 mL (18.0 g, 172. mmol, 3.6 eq) of concentrated hydrochloric acid slowly in 3 mL portions. [Note 2] (Observation 1: melting of nitrobenzene)
3. After stirring for over 30 minutes in the boiling water bath, assemble the reflux apparatus on the flask **A**. (refer to **1.3 Distillation**)
   Reflux for 40 to 50 minutes.
   During reflux, check the pH of the escaping vapor on the top of the condenser using a damp pH paper.
   (Observation 2: starting time of reflux, color change of pH paper)
4. Meanwhile, prepare 50.0 mL of 15.0 M sodium hydroxide solution. [Note 4]
5. Cool down the reaction mixture to room temperature and check the reaction mixture with TLC <the developing solvent is ethyl acetate: hexane = 1:4> [Note 3, 5] (Observation 3: time of reflux and TLC)
   If nitrobenzene remains on TLC, add 5.0 mL of concentrate hydrochloric acid and reflux again.

Date of experiment : **Room temperature :**

Observation

Observation 1 :

Observation 2 :

Observation 3 :

<Work–up>

6. Disassemble the reflux apparatus and place the reaction mixture in the flask A in the boiling water bath.
   Add the prepared 15.0 M sodium hydroxide solution slowly until the reaction mixture reaches pH=9~10 by pH paper. <the developing solvent is ethyl acetate: hexane =1:2 > [Note 6]
   (Observation 4: amount of 15.0 M sodium hydroxide solution, salt formation, color and TLC)

7. Assemble the distillation apparatus on the flask **A**. Distill the reaction mixture using the proper procedures.
   (refer to **1.3 Distillation**) [Note 7] (Observation 5: state of the distillate)

8. To the receiving flask **B** in the distillation apparatus above, add 8.00 g of sodium chloride in order to saturate the aqueous layer and then add 20.0 mL of methylene chloride. [Note 8]

9. Transfer it to the separatory funnel **C** and extract. And then move the organic layer to the separatory funnel **D**. Check the aqueous layer with TLC. If the product remains, extract by adding another 20.0 mL of methylene chloride. (Observation 6: TLC of the aqueous layer)

10. The combined organic layer is washed with brine and then drawn off carefully through the stopcock into a clean 100 mL Erlenmeyer flask **E**. Rinse the stem with 10.0 mL of methylene chloride. Add a small amount of anhydrous magnesium sulfate ($MgSO_4$) to the organic layer in the Erlenmeyer flask until the solution becomes clear and the solid $MgSO_4$ moves freely in the solution.

11. Filter the resulting organic layer by the appropriate suction filtration procedure, which is dealt in **1.5 Recrystallization**.

Observation 4:

Observation 5 :

Observation 6 :

12. Transfer the resulting solution (**filtrate**) to a pre–weighed 100 mL round bottom flask (**tared flask**). Evaporate methylene chloride by using a rotary evaporator. When the solvent is almost evaporated, aniline remain at the bottom of the flask. [Note 9]

13. Check the purity of the product on TLC and calculate the percent yield. [Note 10] (Observation 7: TLC and percent yield)

Observation 7 :

- Theoretical weight of product ____________ g
- Practical weight of product ____________ g
- Percent yield ____________ %

Checked by ______________________

Note 2

1. No definite mechanism is known for the tin reduction. But it has been reported that it undergoes the stepwise reactions with several intermediates. When the reaction condition (Zn/NH_4Cl) is mild, it stops at the step [2] to produce hydroxylamine (R–NHOH).

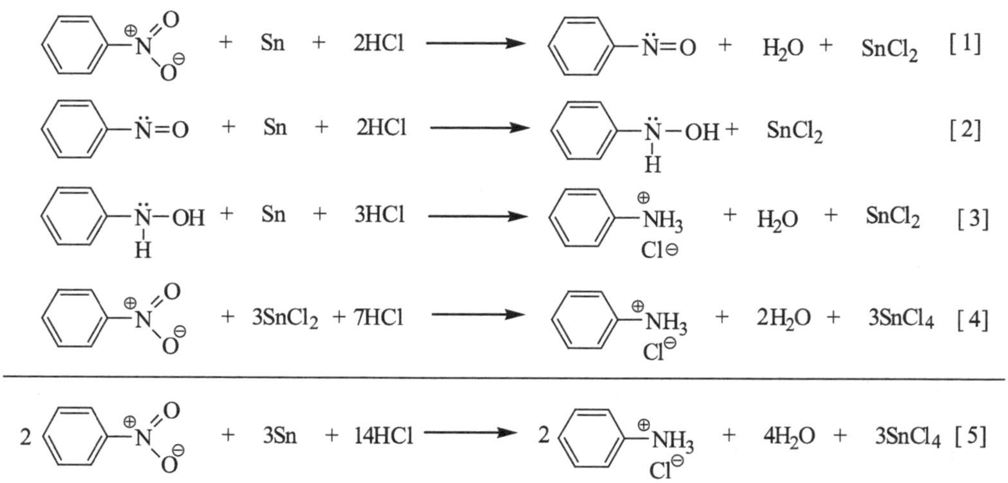

The theoretical mole ratio of aniline and tin is 2:3 since nitrobenzene is reduced not only by tin but also by tin chloride ($SnCl_2$) in step 4. However, two equivalents of tin are used in the actual experiment since the efficiency of the tin reduction is very low.

2. In the step [3], if the reaction mixture is too acidic, *p*–chloroaniline compound may be obtained as a by–product. In order to prevent the reaction mixture from any sudden change of pH, add acid little by little.

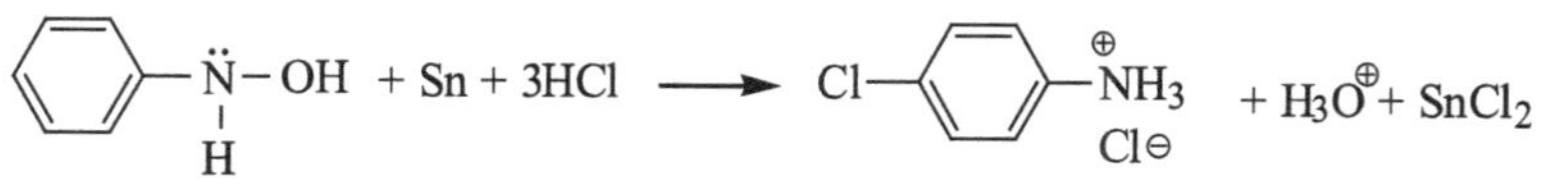

(if too acidic)

3. Check TLC to find out if nitrobenzene has disappeared. If it is remained, reflux for another 15 minutes, and check again.

4. When preparing sodium hydroxide solution, add sodium hydroxide to the flask in an ice bath with stirring.

5. A small amount of remained tin can be ignored as it can easily be removed from the reaction mixture by filtration.

6. When the reduction is complete, the reacting mixture will have the salt forms – $(R\text{–}^{+}NH_3)_2SnCl_4$ or $(R\text{–}^{+}NH_3)_2SnCl_6$ –. Add basic solution to convert them into a primary amine. Add sodium hydroxide slowly since a large amount of heat produced by the acid –base neutralizing reaction may lower the percent yield. In order to prevent a rapid change from a weak acid to a strong base in the reaction mixture, the addition rate of 15.0 M sodium hydroxide solution needs to be well controlled. The TLC pattern of the reaction mixture is not clear until the neutralization is completed.

7. The azeotropic mixture of aniline and water will be distillated at 94~95℃. At the beginning, the distillate will be cloudy or misty since it contains aniline and water. But when the amount of aniline in the distillate decreases, the condensed vapor becomes clear. When the distillate is clear, stop the distillation. About 25~30 mL of distillate will be obtained. (refer to **1.3 Distillation**)

8. The distillate is divided into two layers. The clean lower layer is aniline. The cloudy upper layer contains water and aniline. On adding sodium chloride, the aniline dissolved in water moves to the lower layer. To make an effective extraction of aniline, add methylene chloride. The density of methylene chloride is high, so the organic layer will stay at the bottom. Add enough amount of sodium chloride until it precipitates at the bottom of the flask.

9. The boiling point of methylene chloride is 39℃, and aniline, 184. 0℃, so it can be easily separated.

10. About 65~77 % of percent yield can be obtained. The yield obtained by the evaporated product is higher since methylene chloride is not completely removed by a rotary evaporation.

Keywords for Reduction

Chemical equation for Synthesis of Benzyl alcohol and Aniline

Reaction mechanism for Synthesis of Benzyl alcohol and Aniline

2.6 Nitration

Keywords : nitronium, explosive, Volatile Organic Compound. electrophilic substitution, nucleophilic substitution, Nonlinear Optics

Many nitro compounds react vigorously with oxygen and create explosions. So they are often used for making explosive substances. Explosions produce an instantaneous release of large quantities of heat and hot gases – e.g. N_2, H_2O and CO_2. They set up a tremendous amount of shock waves to generate pressures up to 700,000 atm, which cause enormous physical devastation to the surroundings.

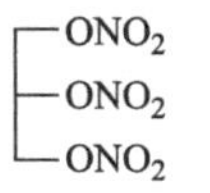

nitroglycerin

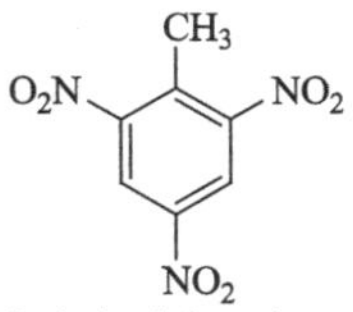

2, 4, 6-trinitrotoluene

(T.N.T)

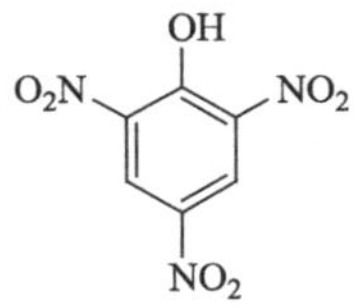

picric acid

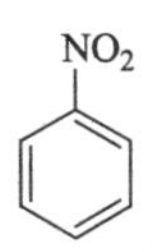

nitrobenzene

Nitromethane (CH_3NO_2, enthalphy of combustion: 747.6 kJ/mol) is used as an engine fuel for racing cars or model planes, which need the explosive power. Some of nitro compounds are used as gunpowder such as nitroglycerin, T.N.T (2,4,6–trinitrotoluene), and picric acid.

Nitrobenzene is a nitro compound but is not as explosive as others. It is mainly used as an extraction solvent in the oil industry or as a solvent for producing shoe polish or dye since it is a good solvent for organic compounds. Nitrobenzene also dissolves many inorganic compounds such as $AlCl_3$, $ZnCl_2$ and is used as a solvent for recrystallization. Recently, the Ministry of Environment regulated nitrobenzene, styrene, toluene, and tetrachloroethylene, which are widely used for manufacturing chemical goods and painting as **Volatile Organic Compounds (V.O.C)**. The acute diseases caused by VOC include a paralysis of the central nervous system (CNS) and irritation of mucous membranes such as eye, skin and the respiratory organs. Due to the new regulation of the chemical industry, nitrobenzene is more often used for dye and perfume than as a solvent in manufacturing processes. Most nitro compounds including nitrobenzene are toxic and easily absorbed through skin and the respiratory organs. Therefore, they should be handled with care.

When the nitro group is introduced in benzoic acid, its pK_a considerably decreases since the nitro group is the strong electron withdrawing group. The pK_a value varies according to the distance between the carboxylate anion and the nitro group as the inductive effect takes effect through the σ–bond.

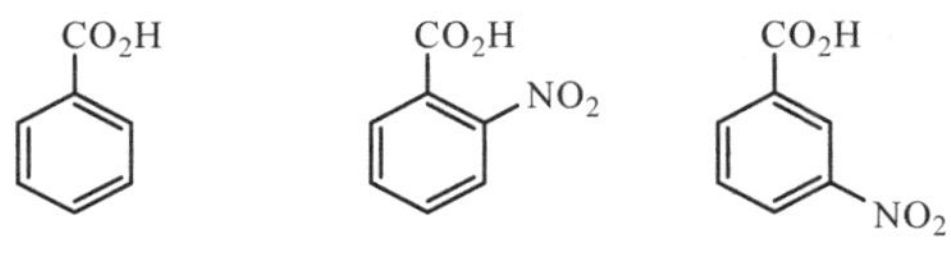

pK_a : 4.20 2.18 3.46

Aromatic nitro compounds containing strong electron withdrawing groups take part in the nucleophilic aromatic substitution. Especially, in the case of 2,4,6–trinitrochlorobenzene, the strong inductive effect and the resonance effect of nitro groups can stabilize the carbanion intermediate. As a result, the hydroxyl group can readily substitute the chloride group.

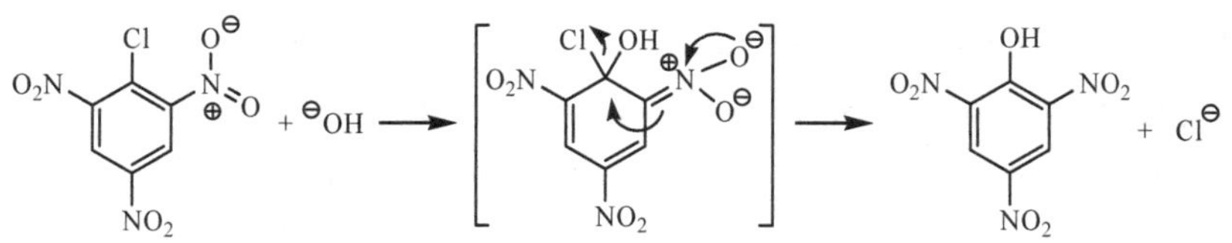

If *p*–nitrophenyl group is added to the amide group of **NOMEX**, the synthesized polymer can be investigated as the candidates for the **nonlinear optics (N.L.O)** material. The electrons are delocalized through the polymer in the solid state by electron withdrawing of the nitro group in the *para* position of the benzene ring and by electron donating of the lone–pair electrons on nitrogen of amide group in the firm polymer chain **(Push–Pull mechanism)**. The resulting dipoles can be arranged in order by electric poling and the synthesized polymer is expected to show nonlinear optical phenomena.

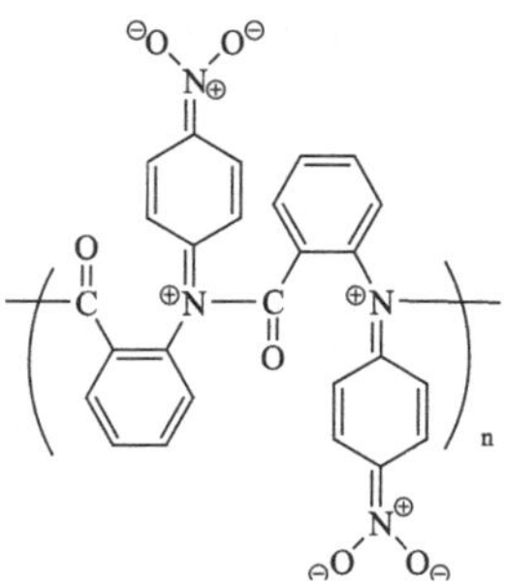

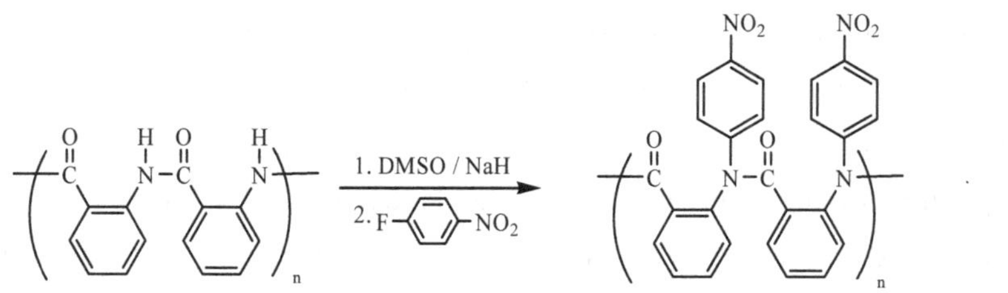

Poly(*m*-pheylene isophthalamide)
NOMEX

N-*p*-notrophenyl poly(*m*-pheylene isophthalamide)

In this experiment, methyl *m*–nitrobenzoate is synthesized by aromatic electrophilic substitution reaction.

Chemical equation

Experiment

Reagents

Name	Molecular formula	Structural formula	M. W.	m.p. (℃)	b.p. (℃)	Density (g/mL)
Methyl benzoate						
Sulfuric acid (H_2SO_4)						
Nitric acid (HNO_3)						
Methyl *m*-nitro benzoate (Product)						

Apparatus

⇨ magnetic stirrer, magnetic spin–bar, lab jack, ice bath, vacuum pump

Glassware

⇨ 100 mL two–neck round bottom flask, 100 mL Erlenmeyer flask, 100 mL graduated cylinder, thermometer, 100 mL beaker, pipette, Büchner funnel, filter flask

Experimental procedure

<Reaction>

1. In a 100 mL two–neck round bottom flask **A**, place 3.0 mL (3.30 g, 24.1 mmol, 1.0 eq) of methyl benzoate in 20.0 mL of hexane and set up a thermometer. [Note 1]
   To monitor the reaction, prepare 3 or 4 TLC plates and spot them with the starting material.
   (Observation 1: reaction time and amount of reagent used.)
2. Prepare an **ice–salt bath using sodium chloride** and cool down the flask **A**. [Note 2] (Observation 2: temperature of ice–salt bath)
3. Carefully add 7.0 mL (12.9 g, 131.0 mmol) of concentrated sulfuric acid to flask **A** and stir well. Be careful that the magnetic spin–bar does not collide with the bulb of thermometer.
   (Observation 3: change in temperature of the flask)
4. In a 100 mL Erlenmeyer flask **B** in an ice bath, prepare the solution of 2.0 mL of concentrated nitric acid and 2.0 mL of concentrated sulfuric acid. [Note 3]
5. Add the cold solution in flask **B** to flask **A** using a pipette. Adjust the addition rate to keep the temperature of the resulting mixture in flask **A** below 15℃. [Note 4, 5]
   (Observation 4: change in temperature)
6. Check the reaction mixture every 20 minutes by the pre–spotted TLC plate.

   <the developing solvent is ethyl acetate : hexane = 1 : 4>
   When methyl benzoate is completely disappeared on TLC, continue stirring for another 10 minutes. (Observation 5: TLC)

Date of experiment : **Room temperature :**

Observation

Observation 1 :

Observation 2 :

Observation 3 :

Observation 4 :

Observation 5 :

<Work–up>

7. Prepare 25.0 g of finely crushed ice in a 100 mL beaker **C**. Transfer the reaction mixture in the flask **A** to the beaker **C** in an ice bath. (Observation 6: change in the solution)

8. When no more crystal is precipitated, filter it by the appropriate suction filtration procedure that is dealt in **1.5 Recrystallization**. Wash the solid with the cold solution made up by mixing 15.0 mL of ice water and 5.0 mL of methanol (3:1).
(Observation 7: color and shape of the crystal)

9. Collect the white crystal. Weigh the crude product before recrystallization. Dissolve a small amount of it in ethyl acetate. Check TLC of the crude product and the mother liquor. [Note 6]
(Observation 8: TLC)

10. Recrystallize the crude product using methyl alcohol by the appropriate procedure that is dealt in **1.5 Recrystallization**.
The amount of methanol required is about twice the weight of the crude product.
(Observation 9: weight, color, state of the crystal before and after the recrystallization.)

11. Weigh the final product, calculate the percent yield, and measure the melting point. [Note 7,8,9]

Observation 6 :

Observation 7 :

Observation 8 :

Observation 9 :

- Theoretical weight of product ____________ g
- Practical weight of product ____________ g
- Percent yield ____________ %

Checked by ______________________

Note

1. In other laboratory textbooks, this reaction is carried out without a solvent. However, we use hexane as a solvent in order to facilitate handling of the reagent and to control the reaction temperature.

2. On adding sodium chloride to the ice can lower the temperature in an ice bath significantly. That is also why calcium chloride is spread over to the snow–covered roads in winter. (refer to **1.5 Recrystallization**)

Mixture of water and ice	:	+ 5~0℃
water : KCl = 4 : 1	:	–12℃
ice : NaCl = 3 : 1	:	–21℃
ice : $CaCl_2$ = 1.2 : 2	:	–39℃
dry ice and acetone	:	–77℃
(indicate parts by weight)		

3. 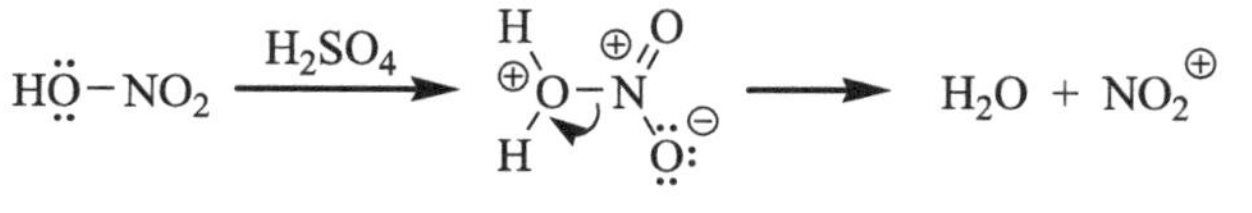

As water is removed during this process, a strong acidic condition is required to produce NO_2^+.

4. Nitration is an exothermic reaction and therefore needs to be cooled down.

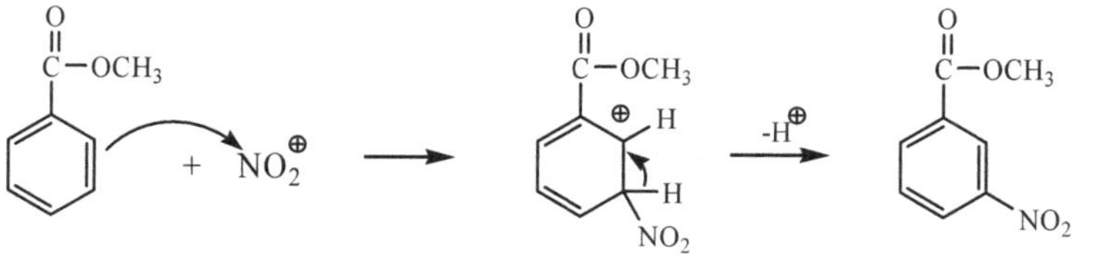

Nitronium is added to the *meta* position where carbocation is formed at the β position of the electron withdrawing carbonyl group.

5. The second nitration of nitrobenzene requires a harsh reaction condition–fuming nitric acid and high temperature–since the nitro group reduces the nucleophilicity of the aromatic compound. The second nitro group is introduced at the *meta* position.

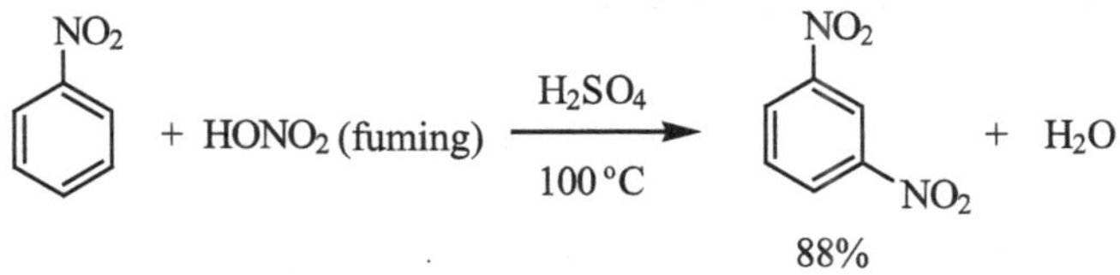

6. Since almost none of the product remains in the mother liquor, it does not affect the total yield.

7. The percent yield is 65~85 %.

8. Nitrobenzens are easily reduced by the various reagents to produce the corresponding anilines.

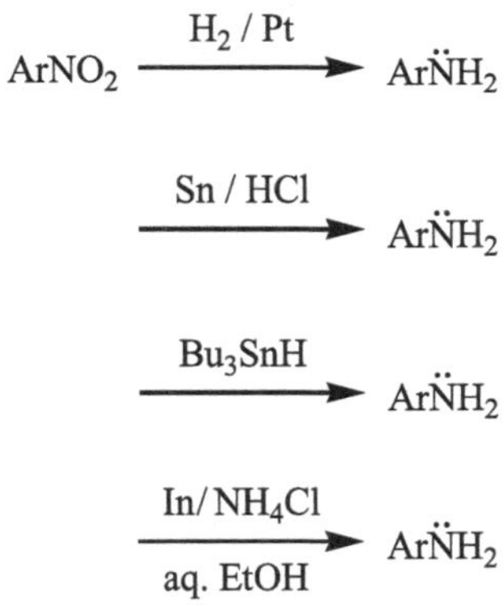

* In = indium

9. The resulting anilines are converted to **diazomium salts**, which are the important starting materials for azo dyes synthesis (refer to **3.1 Synthesis of Azo Dye**)

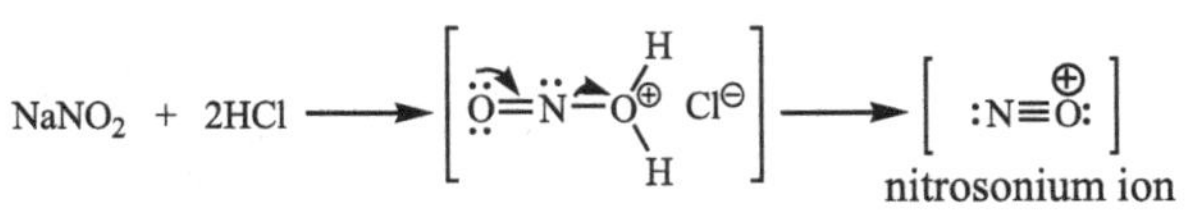

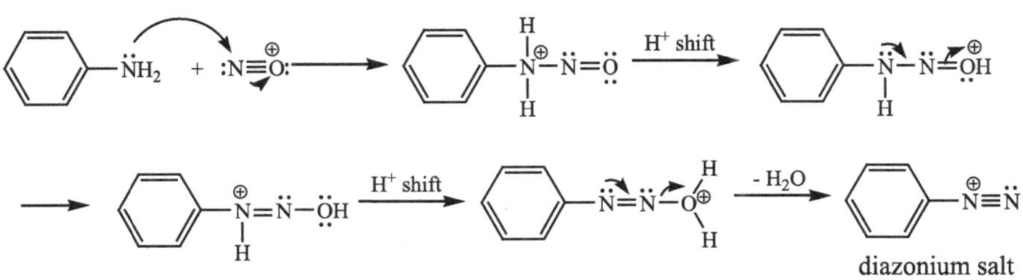

Keywords for Nitro compounds

Chemical equation for Synthesis of Methyl *m*–nitrobenzoate

Reaction mechanism for Synthesis of Methyl *m*–nitrobenzoate

2.7 Cannizzaro Reaction

Keywords : crossed Cannizzaro reaction, hydride, auto redox reaction, Meerwein–Pondorff–Verley reduction, Oppenauer oxidation

Aldehydes which do not hold hydrogen at α carbon of carbonyl groups such as benzaldehyde undergo an auto redox reaction with a strong base. This is called the **Cannizzaro reaction**, which is involved in the **hydride migration**. In the Cannizzaro reaction, formaldehyde gives rise to formic acid and methanol. The formic acid found in ants and other insects is an irritant, which causes itching and swelling after being bitten by them. The dehydrogenase in bacteria decomposes formic acid to carbon dioxide (CO_2), which is consumed by the microorganism.

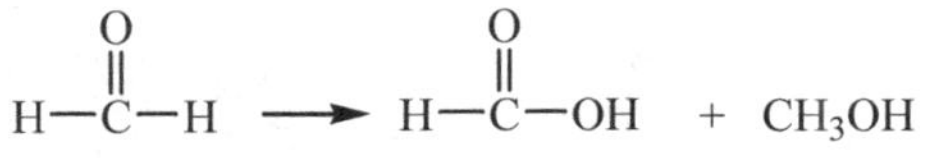

When the oxidant aldehyde differs from the reductant aldehyde, the reaction is called the **crossed Cannizzaro reaction**. When formaldehyde is used in the crossed Cannizzaro reaction, a high yield of alcohol can be obtained with almost any other aldehydes.

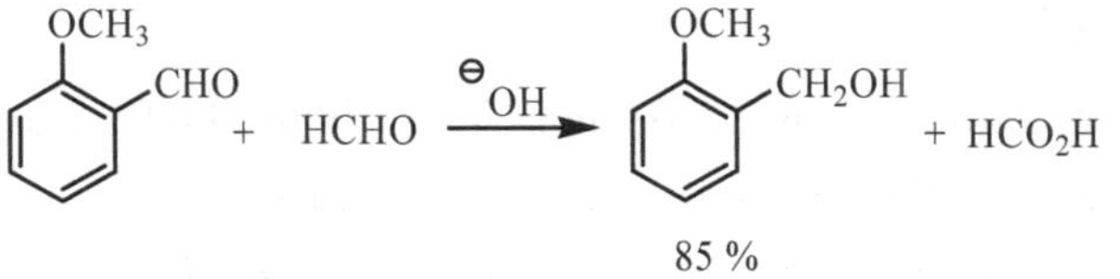

The decomposition of peroxide producing water and oxygen is another good example of auto redox reactions. It proceeds very slowly but reacts violently on the addition of catalyst such as MnO_2.

$$2H_2O_2 \longrightarrow 2\,H_2O + O_2$$

oxygen : -1 -2 0

The hydride plays an important role in the various organic reactions.

1. Internal migration.

The stability of carbocation is in order of primary < secondary < tertiary. The secondary carbocation formed by the initial addition of proton is rearranged to more stable tertiary carbocation by a hydride migration.

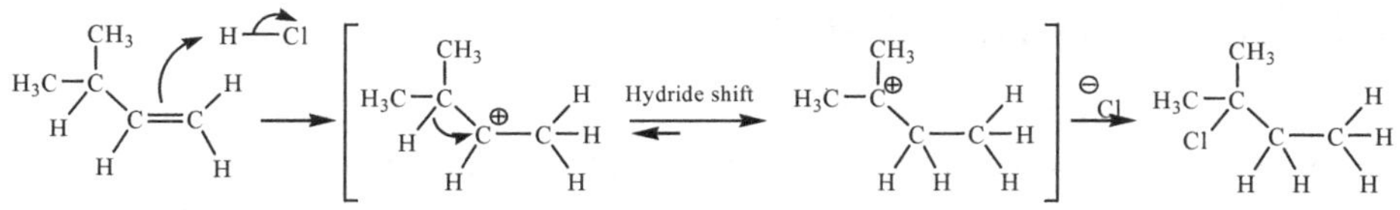

2. Meerwein–Pondorff–Verley reduction

The ketone is reduced to alcohol by aluminum alkoxide through hydride migration. The **driving force** for this reaction is the removal of the volatile product–acetone–.

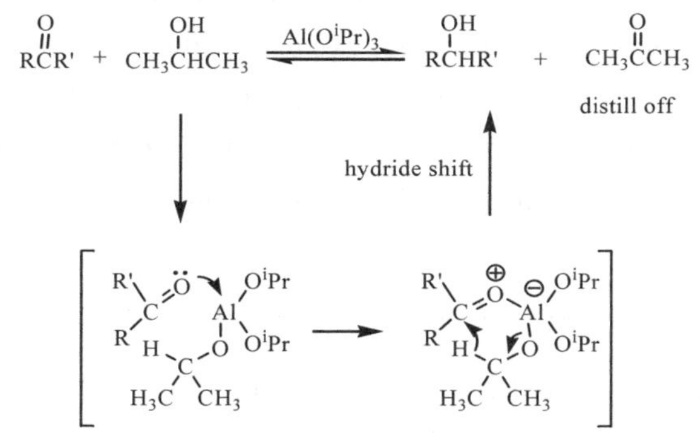

It is the reverse reaction of the **Oppenauer oxidation**, which is used to oxidize secondary alcohols to ketones.

3. Bio oxidation

In the catabolic pathway of fats in a living body, the hydride is involved as in the crossed Cannizzaro reaction. **NAD$^+$** is one of biochemical oxidizing agents for converting alcohols to ketones or aldehydes.

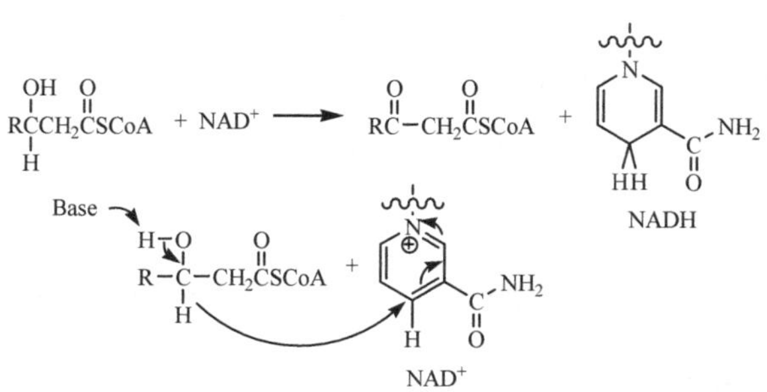

CoA : Co-enzyme A
NAD$^+$: Nicotinamide adenine dinucleotide
NADH : Reduced Nicotinamide adenine dinucleotide

Benzoic acid and benzyl alcohol are synthesized from benzaldehyde under a strong basic condition.

Chemical equation

Experiment

Reagents

Name	Molecular formula	Structural formula	M. W.	m.p. (℃)	b.p. (℃)	Density (g/mL)
Benzaldehyde						
Benzoic acid (Product)						
Benzyl alcohol (Product)						

Apparatus

⇨ magnetic stirrer, magnetic spin–bar, lab jack, UV lamp, TLC plate, ice bath, vacuum pump, evaporator, melting point apparatus

Glassware

⇨ TLC chamber, 50 mL Erlenmeyer flask
100 mL graduated cylinder, 250 mL separatory funnel, 100 mL Erlenmeyer flask, 100 mL one–neck round bottom flask, pH paper, filter flask, Büchner funnel

Experimental procedure

* This reaction takes more than 8 hours. It is recommended that procedure **1** through **3** should be carried out a day before the experiment.

<Reaction>

1. Prepare a 50 mL Erlenmeyer flask **A** in an ice bath, place 3.30 g (58.8 mmol, 3.0 eq) of KOH, 5.0 mL (5.50 g, 70.5 mmol, 3.5 eq) of dimethyl sulfoxide and 10.0 mL of water and stir well. [Note 1]
2. Place 2.0 mL (2.10 g, 19.7 mmol, 1.0 eq) of benzaldehyde and a magnetic spin–bar in a 50 mL Erlenmeyer flask **B**. [Note 2]
3. Add the solution in flask **A** to flask **B** slowly with stirring. When the addition is finished, put an appropriate stopper to seal the flask tightly to prevent benzaldehyde from evaporating. Stir thoroughly.
   To monitor the reaction, prepare 3 or 4 TLC plates and spot them with the starting material. (Observation 1: change in color)
4. Check the reaction mixture every two hours by the pre–spotted TLC plate. <the developing solvent is ethyl acetate : hexane = 1:4, R_f of benzaldehyde = 0.71> (Observation 2 : TLC)
   When benzaldehyde is completely disappeared on TLC, continue stirring for another one hour.

<Work–up>

5. Transfer the reaction mixture to a 100 mL Erlenmeyer flask. And add 50.0 mL of water or more to the reaction mixture until the solid— potassium benzoate— dissolves completely. [Note 3]
6. Transfer the solution to the 250 mL separatory funnel and extract with 20.0 mL of diethyl ether. Remove the aqueous layer and keep it in the flask **C** to recover the benzoic acid later on.
7. Check if benzoic acid or benzaldehyde remains in the organic layer by TLC. <the developing solvent is ethyl acetate: hexane = 1:2> (Observation 3: TLC)

Date of experiment : **Room temperature :**

Observation

Observation 1 :

Observation 2 :

Observation 3 :

8. If benzaldehyde remains, extract with 5.0 mL of 20 % sodium bisulfite solution and check again. [Note 4]
9. If benzoic acid remains, extract with sat'd $NaHCO_3$ and check again.

*** Refer to 1.4 Extraction for the following procedures.**

<Isolation of benzyl alcohol>

10. The organic layer is washed with 20.0 mL of water followed by brine and then is poured through the mouth of the separatory funnel into a clean 100 mL Erlenmeyer flask. Rinse the empty separatory funnel with 10.0 mL of diethyl ether. Add a small amount of anhydrous sodium sulfate (Na_2SO_4) to the organic layer in the Erlenmeyer flask until the solution becomes clear and the solid Na_2SO_4 moves freely in the solution.
11. The resulting organic layer is filtered according to the appropriate suction filtration procedure, which is dealt in **1.5 Recrystallization**.
12. Transfer the resulting solution (**filtrate**) to a pre–weighed 100 mL one–neck round bottom flask (**tared flask**). Evaporate diethyl ether by using a rotary evaporator. When the solvent is almost evaporated, benzyl alcohol will remain at the bottom of the flask.
13. Weigh the flask and check the purity of the product by TLC. <the development solvent is ethyl acetate: Hexane = 1:4> Calculate the percent yield. [Note 5]
(Observation 4: TLC, weight and percent yield)

<Isolation of benzoic acid>

14. Keep adding 6.0 M hydrochloric acid by small amounts to the flask **C** in an ice bath until the pH of the solution becomes 1 on the pH paper. When acid is added, benzoic acid will precipitate to give a white–colored crystal since it is sparingly soluble in water. [Note 6] (Observation 5: formation of a white crystal)

Observation 4 :

Observation 5 :

15. Collect the solid by filtering with suction on the Büchner funnel, wash twice with 10 mL of cold water. Allow it to dry for few minutes and spread it on a clean pre–weighed filter paper to dry in the air.
16. Check the purity of the formed benzoic acid by TLC. (Observation 6: TLC)
17. Weigh the filter paper when it is completely dry. Calculate the recovered amount of benzoic acid and the percent yield. Measure the melting point and calculate the yield of the obtained benzoic acid and compare it with the literature value. [Note 7] (Observation 7: percent yield and m.p.)
18. For further interest, take a small amount of the obtained benzoic acid and recrystallize it in water to get the pure product. (Observation 8: used amount of water)

Observation 6 :

Observation 7 :

Observation 8 :

- Theoretical weight of product ____________ g
- Practical weight of product ____________ g
- Percent Yield ____________ %

Checked by ______________________

Note

1. Cannizzaro reaction usually takes more than 24 hours. The addition of DMSO to the reaction medium enables the reaction to be completed within 10 hours.
   Dimethyl sulfoxide (DMSO), dimethylformamide (DMF), hexamethyl phosphoramide (HMPA), acetone and acetonitrile are polar aprotic solvents.

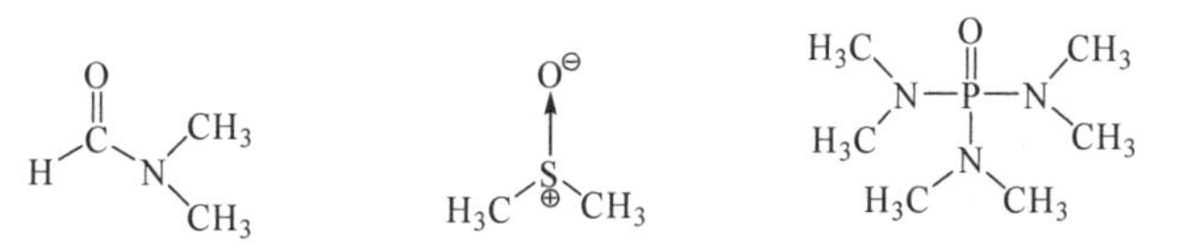

	DMF	DMSO	HMPA
b.p. (℃)	153	189	232
dielectric constant (ε)	36.7	49	30

   These can be used as solvents for organic reactions involving ionic reagents or ionic reaction intermediates since they dissolve many salts. They tend to strongly solvated metal cations and generate the **naked or free anions**, which have a very strong nucleophilicity. (refer to **2.1 Esterificaiton**)

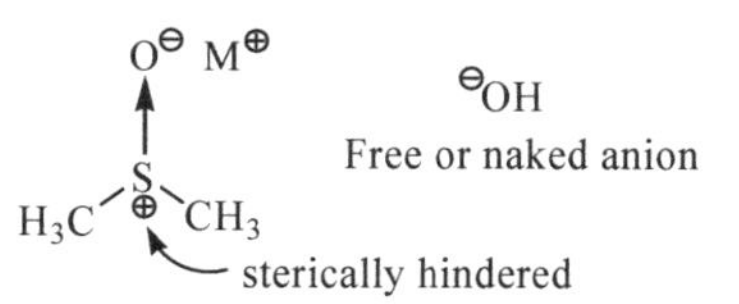

DMSO

   It is reported that the rates of nucleophilic substitution reactions in these solvents are up to 200,000 times higher than those in alcohols.

2. An addition of hydroxide to benzaldehyde gives rise to carboxylic acid and alkoxide. The alkoxide readily deprotonates the resulting carboxylic acid to produce a more stable carboxylate anion and the corresponding alcohol. The **formation of the stable carboxylate** is the **driving force** for the Cannizzaro reaction. The formation of the stable anion can be explained by the difference in pK_a of the carboxylic acid and the alcohol. (refer to **1.4 Extraction**) In additon, the Canizzarro reaction essentially goes to completion since the benzyl alcohol is removed from the reaction mixture as a precipitate.

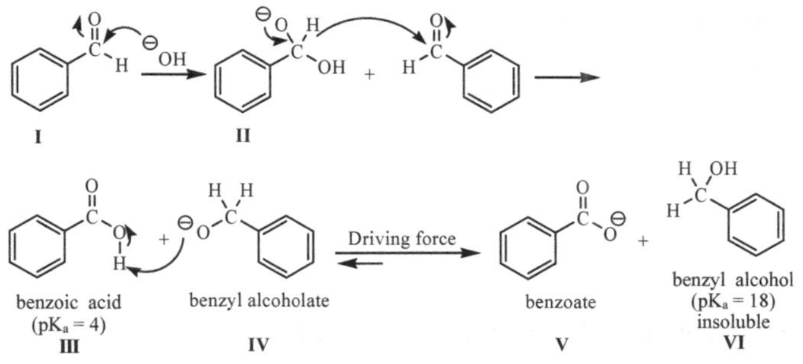

3. Using DMSO increases the reactivity and reduces the reaction time. However, it causes a problem in isolating the desired product from the reaction mixture since it dissolves well in both organic and inorganic compounds. To make the isolation easier, add 10 times or more amount of water to the reaction mixture to lower the solubility of the organic product in DMSO. The organic product usually precipitates as a solid because of its low solubility in water.

4. In order to remove the remaining benzaldehyde, 20 % sodium bisulfite is added. Bisulfite attacks the carbonyl group in aldehyde to form a sulfonate ion, which is soluble in water and removed by washing with water.
   <R_f of benzaldehyde = 0.71 and R_f of benzyl alcohol = 0.77>

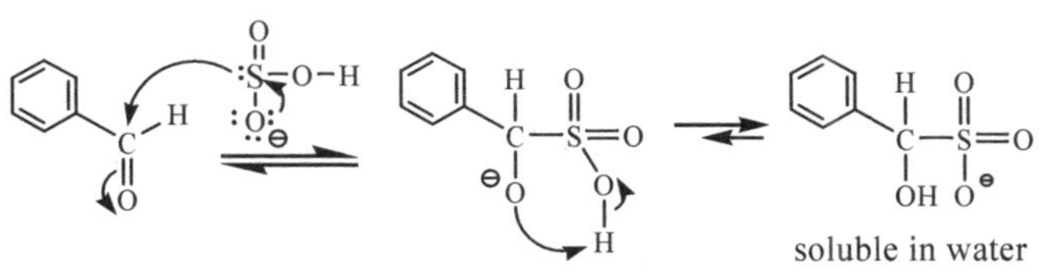

5. The percent yield of benzyl alcohol is 78~90 %.

6. When hydrochloric acid is added to benzoate ion, it is converted into benzoic acid, which precipitates in water.

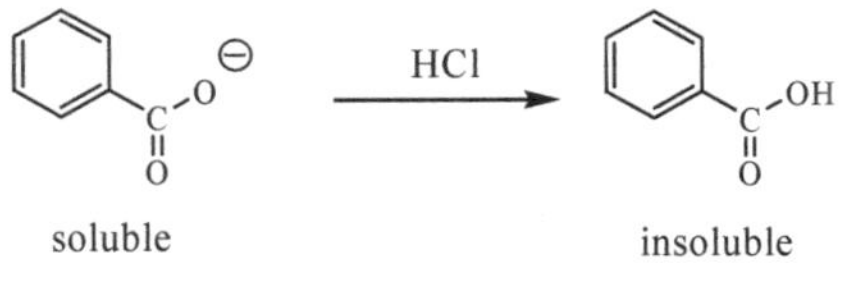

7. The percent yield of benzoic acid is 74~80 % and melting point is 114.0~116.0℃. (lit. value: 122.4℃)

Keywords for Cannizzaro reaction

Chemical equation for Cannizzaro reaction of Benzaldehyde

Reaction mechanism for Cannizzaro reaction of Benzaldehyde

2.8 Aldol Condensation

Keywords : *E,Z*–enolate, chair–like transition state, 1,3–diaxial interaction, 2,3–allylic interaction, retro–aldol condensation

Organic reactions can be divided into the **functional group interconversion** and the **C–C bond formation**. For example, the functional group interconversion is a reaction where an alcohol is oxidized to a carboxylic acid. It is used to transform the property or reactivity of a organic substance. The C–C bond formation is used to synthesize a big molecule from small molecules and provides a basic synthetic method for building the carbon skeleton.

The aldol condensation is the typical C–C bond formation through the enolate intermediate. A new C–C bond is formed with one carbonyl compound acting as a **nucleophile** and the other as an **electrophile**.

LDA (lithium diisopropyl amide) is a strong base, which is frequently used to generate an enolate from a carbonyl compound. The formation of enolate with a base undergoes a **6–membered transition state**. In this transition state, the enolate structure is determined by the starting material and the base. Usually, (*E*)–enolate is formed as the major product because of the **1,3–diaxial interaction** between the substituents at the α postion of the carbonyl group and the isopropyl group from LDA.

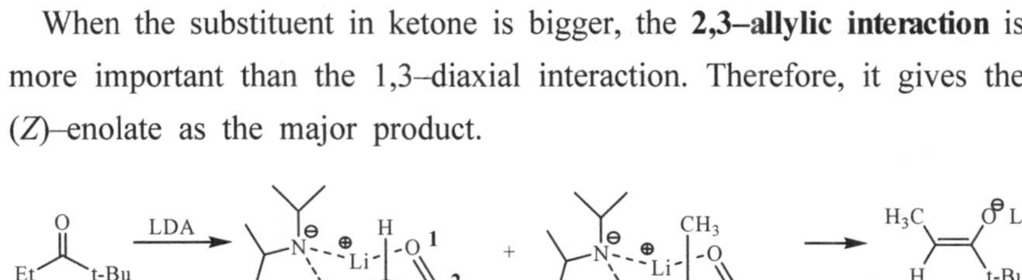

When the substituent in ketone is bigger, the **2,3–allylic interaction** is more important than the 1,3–diaxial interaction. Therefore, it gives the (*Z*)–enolate as the major product.

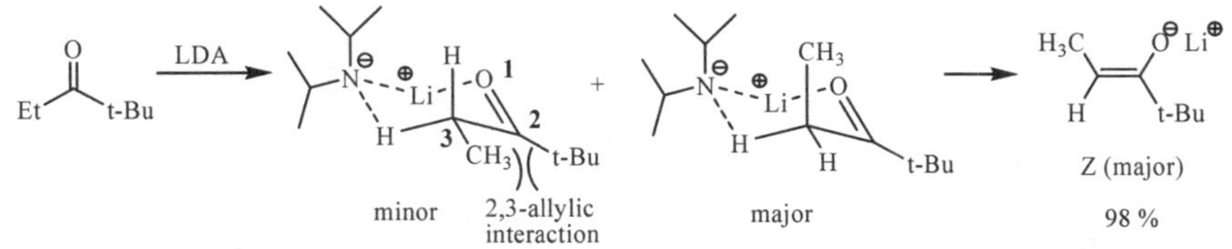

When the bulky base is used, the 1,3–diaxial interaction becomes predominant, giving a (*E*)–enolate as the major product.

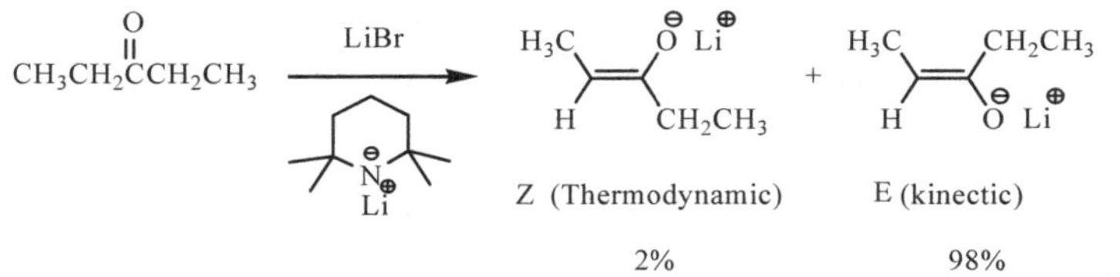

The aldol condensation is also carried out through a 6–membered transition state. R^3 is located at the equatorial position in a **chair–like transition state** since it is bigger than hydrogen in volume The (*Z*)–enolate produces *thero* adduct and the (*E*)–enolate produces *erythro* adduct, respectively. The substituents in the same direction are called ***threo(syn)***, and the substituents in another direction are called ***erythro(anti)***.

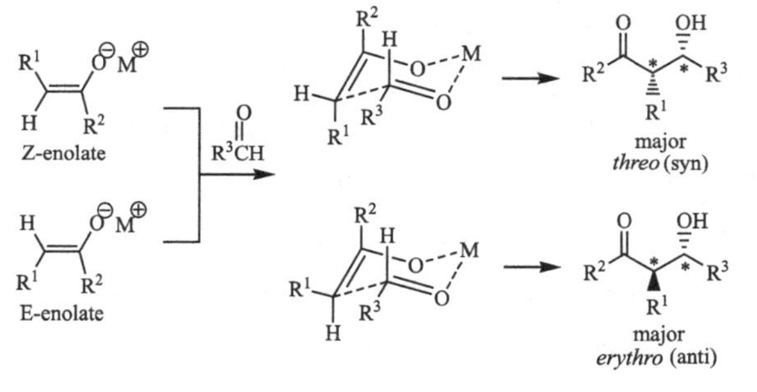

The aldol condensation is a reversible reaction. When a base is acted upon the initially formed *β*–hydroxy compound, a reverse reaction occurs to restore the starting materials. This is called the **retro–aldol reaction.**

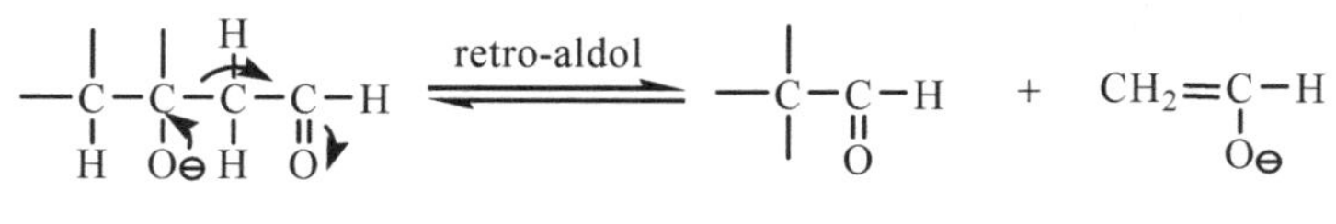

The retro–aldol reaction plays an important role in breaking down a big biomolecule into small molecules.

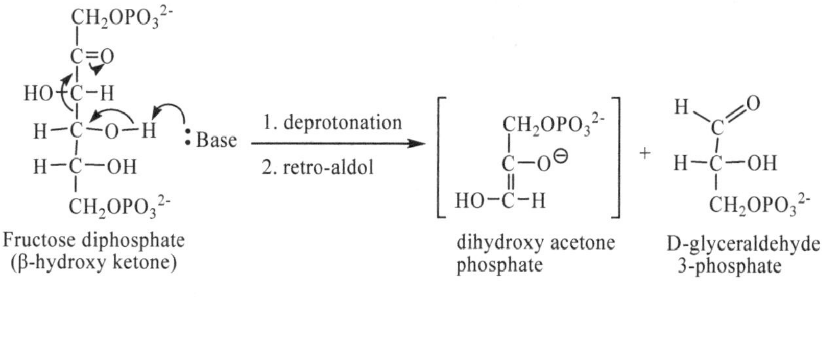

Chemical equation

Experiment

Reagents

Name	Molecular formula	Structural formula	M. W.	m.p. (℃)	b.p. (℃)	Density (g/mL)
Benzaldehyde						
trans,trans–Dibenzylidene–acetone (Product)						

Apparatus

⇨ magnetic stirrer, magnetic spin–bar, lab jack, UV lamp, ice bath, evaporator, vacuum pump

Glassware

⇨ 100 mL one–neck round bottom flask, 250 mL one–neck round bottom flask, 25 mL Erlenmeyer flask, 100 mL graduated cylinder, 250 mL separatory funnel, TLC chamber, filter flask, Büchner funnel

Experimental procedure

<Reaction>

1. Prepare a 250 mL one–neck round bottom flask **A** in an ice bath, place 40.0 mL of ethanol and 50.0 mL of water. [Note 1]
   Add 10.0 g (250 mmol) of sodium hydroxide (NaOH).
   (Observation 1 : change in temperature of solution)

2. Prepare the solution of 5.0 mL (5.20 g, 49.2 mmol, 1.8 eq) of benzaldehyde and 2.0 mL (1.60 g, 27.2 mmol, 1.0 eq) of acetone in a 25 mL Erlenmeyer flask **B**. [Note 2] (Observation 2 : time)
   To monitor the reaction, prepare 3 or 4 TLC plates and spot them with the starting material.

3. Slowly add half of the solution in the flask **B** to the flask **A**. Adjust the addition rate to keep the temperature of the resulting mixture in flask A below 2℃. [Note 3]
   Stir vigorously for 15 minutes.
   (Observation 3: temperature and state of the solution)

4. Add the other half of the solution in the flask **B** to the flask **A** in the same manner in procedure 3.
   (Observation 4: time of addition, change in temperature of solution)

5. Check the reaction mixture every 10 minutes with the pre–spotted TLC plate. <the developing solvent is ethyl acetate: hexane = 1 : 3>
   When benzaldehyde is completely disappeared on TLC, continue stirring for another 10 minutes. [Note 4]
   (Observation 5: time, TLC, shape and color of the precipitate)

Date of experiment : **Room temperature :**

Observation

Observation 1 :

Observation 2 :

Observation 3 :

Observation 4 :

Observation 5 :

<Work–up>

6. Filter the reaction mixture in the flask A by the appropriate suction filtration procedure that is dealt in **1.5 Recrystallization**. Wash the precipitate twice with 50.0 mL of cold water. [Note 5] (Observation 6: color after washing with water)

7. Dissolve the collected solid compounds in 100.0 mL of ethyl acetate. [Note 6]

8. Transfer the resulting solution to a 250 mL separatory funnel and separate the aqueous layer. Check TLC on both layers.

9. If benzaldehyde remains in the organic layer, remove it by washing with 20.0 mL of 20% sodium bisulfate solution. (refer to **2.7 Cannizzaro reaction Note 4**)

10. The organic layer is washed with brine and then poured through the mouth of the separatory funnel into a clean 250 mL Erlenmeyer flask. Rinse the empty separatory funnel with 10.0 mL of ethyl acetate. Add a small amount of anhydrous magnesium sulfate ($MgSO_4$) to the organic layer in the Erlenmeyer flask until the solution becomes clear and the solid $MgSO_4$ moves freely in the solution. (Observation 7: separation of layers)

11. Filter the resulting organic layer in the appropriate suction filtration procedure, which is dealt in **1.5 Recrystallization**.

12. Transfer the resulting solution (**filtrate**) to a pre–weighed 100 mL one–neck round bottom flask (**tared flask**). Evaporate diethyl ether by using a rotary evaporator. When the solvent is almost evaporated, *trans*, *trans*–dibenzylideneacetone will remain at the bottom of the flask. (Observation 8: weight of the solid)

Observation 6 :

Observation 7 :

Observation 8 :

13. Check the purity and the melting point of the crude product using TLC. [Note 7] (Observation 9: TLC and melting point). The crude product should be pure enough to use in **3.3 Nazarov reaction**.

14. Calculate the percent yield of the solid compound obtained. (Observation 10: percent yield)

15. Take 1.00 g of the crude product and recrystallize it with 15.0 mL of ethanol. Calculate the percent yield of the recrystallized product. [Note 8, 9] (Observation 11: used volume of ethanol, percent yield)

Observation 9 :

Observation 10 :

Observation 11 :

- Theoretical weight of product ____________ g
- Practical weight of product ____________ g
- Percent yield ____________ %

Checked by ______________________

Note

1. Ethanol is used as a co–solvent to dissolve both benzaldehyde and benzal acetone which is the key intermediate in this reaction.

2. Although the theoretical amount of benzaldehyde is 2.0 equivalent of acetone, a less amount is used here to cut down the process of removing unreacted benzaldehyde from the reaction mixture.

3. Sodium hydroxide is used to generate an **enolate** from acetone. Because water is more acidic than acetone, only a small amount of enolate is formed at the beginning. The resulting enolate is added to benzaldehyde to produce a *β*–hydroxy compound. Removal of water from the *β*–hydroxy compound produces a stable **enone** (*α*, *β*–unsaturated carbonyl compound). It is the **driving force** for this reaction. As a result, the enolate is continuously consumed as the reaction proceeds and the equilibrium shifts to produce the enolate from acetone by **Le Châtelier's principle**.

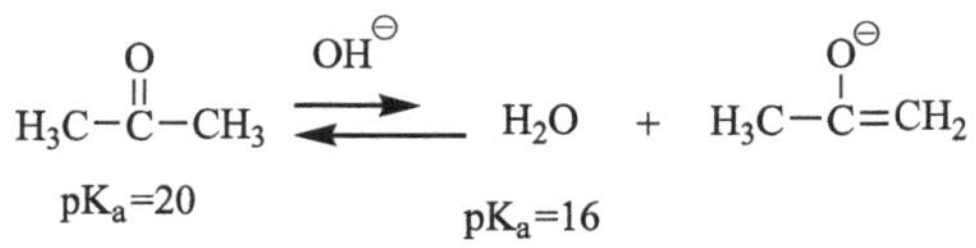

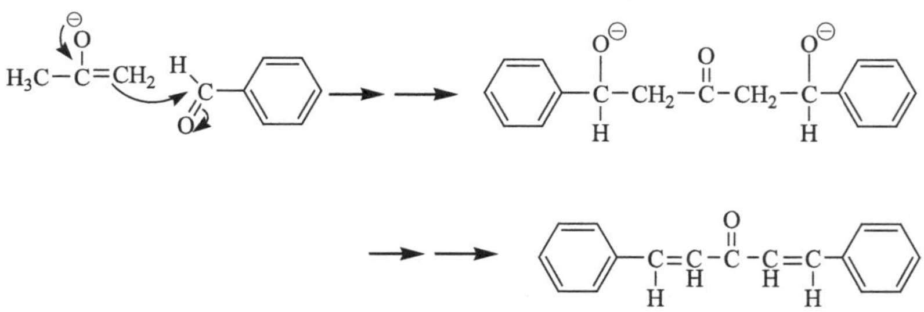

4. R_f of benzaldehyde is 0.68. The formation of *β*–hydroxy carbonyl compound in aldol condensation is **reversible**. But the dehydration step produce an enone is **irreversible**.

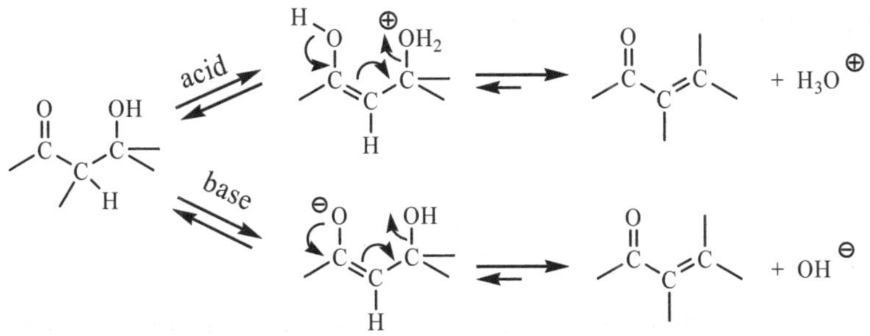

5. When washing the precipitate with water, decrease reduced pressure for a moment. The water is poured and the precipitate is stirred thoroughly with a glass rod followed by reapplying reduced pressure in full strength. As the resulting product is not soluble in water, wash it with a sufficient amount of water to remove any sodium hydroxide remained.

6. The vacuum filtered compound contains water even after drying, and the percent yield exceeds 100 %. When it is dissolved in ethyl acetate, water contained in the product is separated as the aqueous layer.

7. When vacuum filtered, a yellow–colored solid is collected. This should be pure enough to be used in the next reaction.

8. *trans*, *trans*–Dibenzylideneacetone(dibenzal acetone) is not toxic and used to block sunlight. The percent yield is 75~85 %.
m.p. before recrystallization : 103.7~108.2℃
m.p. after recrystallization : 102.7~106.3℃ (lit.value: 104.0~107.0℃)

^{1}H NMR $CDCl_3$: δ7.14 (d, J=12.1, 2H^{α}),
δ 7.55 (m, 10H, aromatic), δ 7.75(d, J=12.1, 2H^{β})

9. Hydrogens at the α–position and at the β–position of enone show distinctive chemical shifts as below.

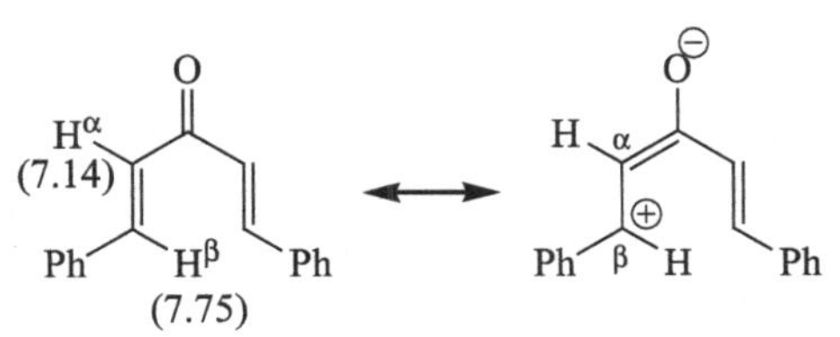

β–Proton is shifted more down field than α–proton because its electron density is decreased by the resonance structure of enone.

Keywords for Aldol condensation

Chemical equation for Synthesis of *trans,trans*–Dibenzylideneacetone

Reaction mechanism for Synthesis of *trans,trans*–Dibenzylideneacetone

CHAPTER

03

Intermediate Organic Reactions

3.1 Synthesis of Azo Dye

Keywords : chromophore, chromogen, auxochrom, diazonium salt, azo compound, light recording material

Only a small quantity of dye is required to color a large amount of fabric or organic substance. A dye molecule consists of a **chromogen**, a **chromophore**, and an **auxochrom**.

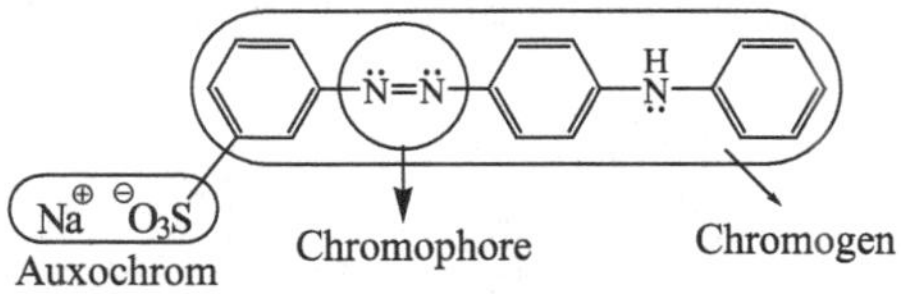

Metanil Yellow

Chromogen is the part of a dye which includes a chromophore and an aromatic or hetero aromatic group.

Chromophore (–N=N–, –N=O, =C=O, –C=C–) has a multiple bond between nitrogen, oxygen, and/or carbon atom. Auxochrom (–NR_2, –OH, –COOH, –SO_3H–) is the part that can be ionized by an acid or a base, which gives the fabric a unique color and a property of fastness. Chromophore itself is a light–colored and has a low quality of fastness, so the combined properties of a chromophore and an auxochrom determine the color and the fastness of a dye.

The most important and popular synthetic dyes are the **azo dyes**. An electron donating group on the aromatic ring is conjugated with an electron withdrawing group on the other ring in azo dyes. They produce a wide spectrum of colors such as red, orange, yellow, blue and brown. And they have distinct deep colors.

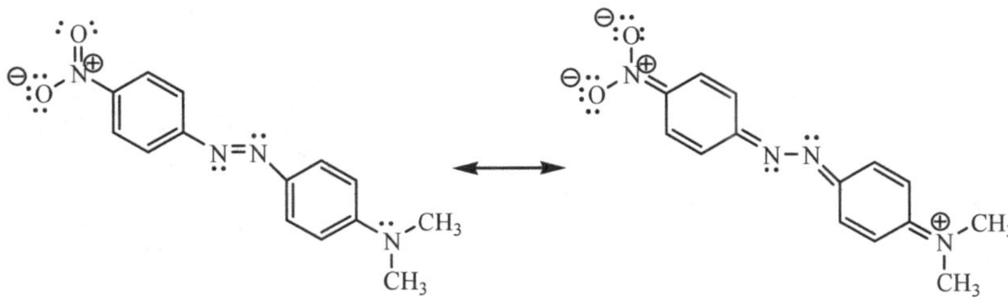

4-dimethylamino-4'-nitrobenzene
λ_{max} 478nm (ε 33,110)

Many azo dyes have a sulfonate groups, which enable them to adhere firmly to the fibers such as silk and wool.

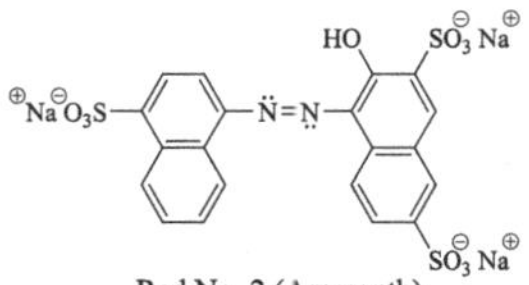

Red No. 2 (Amaranth)

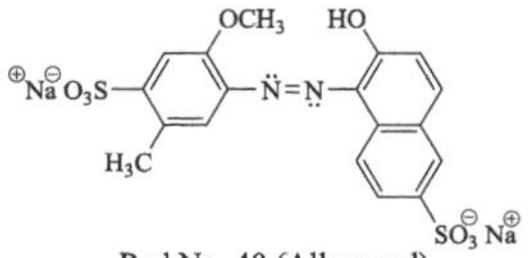

Red No. 40 (Allura red)

The azo dyes are also major food dyes. Before the ban in the United States, Red Number 2 was the most widely used food dye in the food industry, and it was applied to everything from ice cream to cherry soda. Red Number 40, the most recently accepted food dye was developed for the replacement of Red Number 2.

An azo compound with a phenyl group on both ends of the azo linkage can have *cis* and *trans* isomers just as the carbon–carbon double bond.

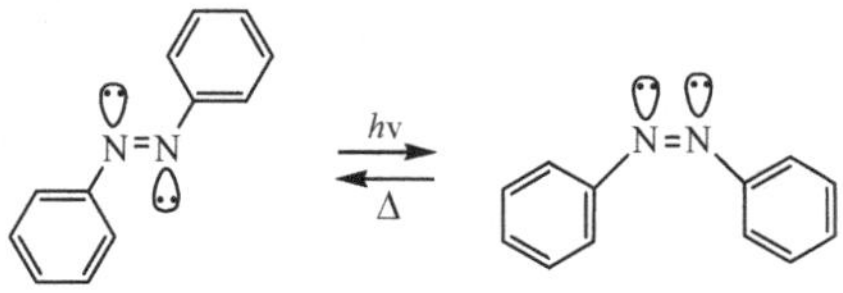

I II

The dyes used on CD or DVD are the cyanine type, the phthalo cyanine type, or the azo type.

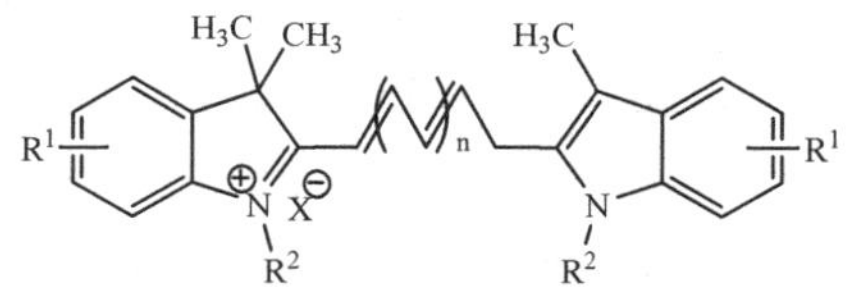

Cyanine type

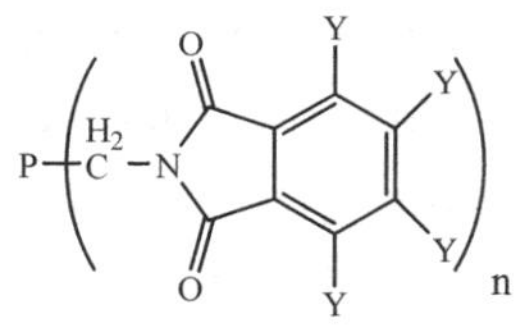

Phthalocyanine type

The blue CD made from the azo type dye is more widely used than other types for recording due to its durability and cost efficiency. Especially the **metal azo dye**, which contains a metal atom, is most often used. The metal atom increases the durability and the quality of recording.

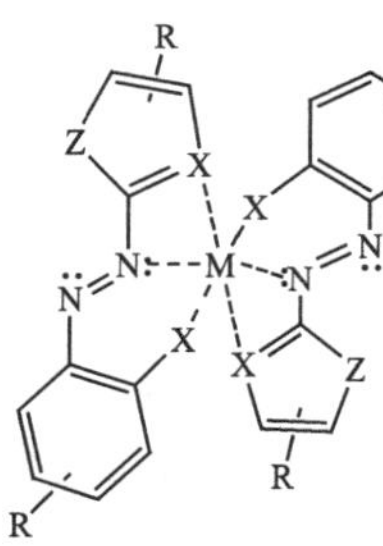

M= Co(III), Ru(III) etc.

X = C, O, S, NH, NR

Z = N, C=C, S

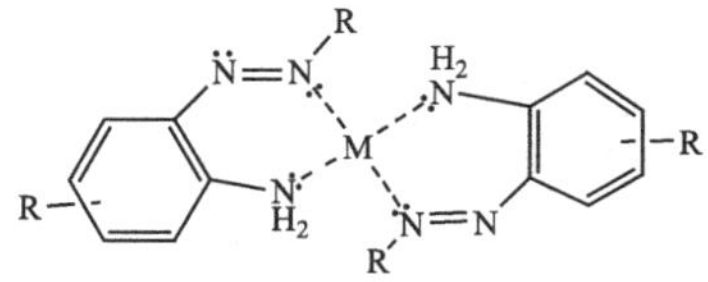

M= Co(II), Ni(II), Fe(II), Pd(II), ZN(II) etc.

In this experiment, the diazonium salt is synthesized and then transformed further into methyl orange or orange II.

Chemical equation

Experiment I

Reagents

Name	Molecular formula	Structural formula	M. W.	m.p. (℃)	b.p. (℃)	Density (g/mL)
Sodium carbonate						
Sulfanilic acid						
Sodium nitrite						
N,N-Dimethylaniline						
Acetic acid						
Methyl orange (Product)						

Apparatus

⇨ magnetic stirrer, magnetic spin–bar, lab jack, ice bath

Glassware

⇨ 250 mL Erlenmeyer flask, 50 mL Erlenmeyer flask, 100 mL graduated cylinder, 250 mL beaker, Büchner funnel, filter flask, test tube

Experimental procedure

Because this reaction deals with the salt type of reaction intermediates, it is difficult to monitor the reaction with TLC. Therefore, observing the color change of the reaction mixture is required for monitoring the reaction.

<Making diazonium salt>

1. In a 250 mL Erlenmeyer flask **A**, place 100.0 mL of distilled water.
   Add 275. mg (2.6 mmol, 0.47 eq) of sodium carbonate and 1.00 g (5.8 mmol, 1.1 eq) of sulfanilic acid and heat the reaction mixture with stirring until it dissolves completely. [Note 1]
2. When the reaction mixture dissolves completely, cool it down to room temperature and then place it in an ice bath. [Note 2]
3. Add 375. mg (5.4 mmol, 0.98 eq) of sodium nitrite. [Note 3] (Observation 1: change in temperature)
4. Add 1.0 mL (1.20 g, 2.0 mmol, 2.1 eq) of concentrated hydrochloric acid to 10.0 g of ice in a 50 mL Erlenmeyer flask **B**. Add ice–cold hydrochloric acid in the flask **B** to the flask **A** slowly with a disposable pipette. [Note 4]
5. Stir the resulting mixture thoroughly in the ice bath until the diazomium salt is formed. [Note 5] (Observation 2: change in color)

<Synthesis of Methyl orange>

6. In a test tube, mix 0.7 mL (0.67 g, 5.5 mmol, 1.0 eq) of N,N–dimethylaniline and 0.5 mL (0.52 g, 8.7 mmol, 1.6 eq) of acetic acid. Add the mixture to the flask **A** from the procedure 5. [Note 6] (Observation 3: change in color)
7. Stir the reaction mixture for about 10~15 minutes. [Note 7]

Date of experiment : **Room temperature :**

Observation

Observation 1 :

Observation 2 :

Observation 3 :

<Work–up>

8. Add 10 % sodium hydroxide solution (2.78 M) to the flask **A** until the pH of the reaction mixture becomes 8~9. [Note 8]
   (Observation 4: amount of the 10 % sodium hydroxide solution)
9. Heat the resulting reaction mixture until the precipitate dissolves completely and then add about 10.0 g of sodium chloride. [Note 9]
10. Cool down the reaction mixture to room temperature and then place it in an ice bath to crystallize the product.
11. When no more crystal is precipitated, filter it by the appropriate suction filtration procedure that is dealt in **1.5 Recrystallization** and wash it with 20.0 mL of brine. [Note 10]
    (Observation5: weight of filter paper)
12. After the product is sufficiently dried in air, weigh the product and calculate percent yield. [Note 11, 12]
    (Observation 6: weight, color, and percent yield)

Observation 4 :

Observation 5 :

Observation 6 :

- Theoretical weight of product ____________ g
- Practical weight of product ____________ g
- Percent yield ____________ %

Checked by ______________________

Note

1. Gently heat the reaction mixture just enough to dissolve it completely.

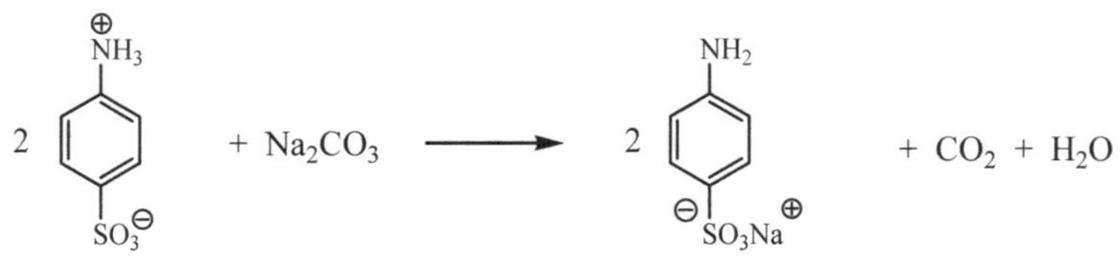

2. While dissolving it, prepare an ice bath in advance.

3. Adjust the amount of sodium nitrite so that the temperature of the reaction mixture does not exceed 10℃.

4. If the temperature rises rapidly, add ice directly to the reaction mixture to control the temperature.

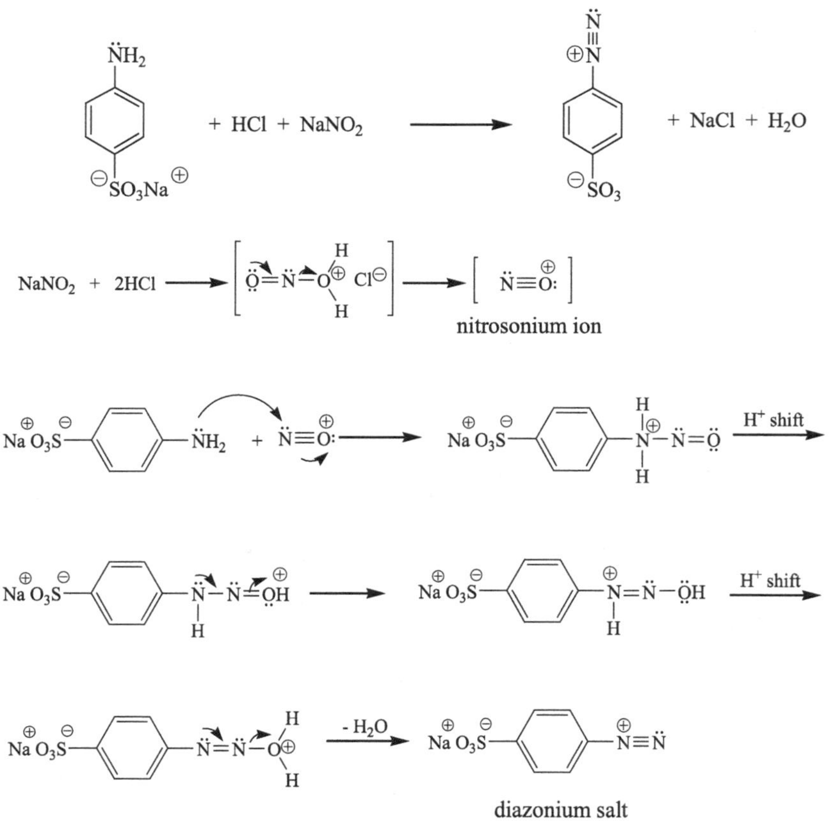

5. The color of the reaction mixture will turn to yellow due to the formation of a diazonium salt. The diazonium salt should be kept below 10℃. If the temperature is higher, it might be hydrolyzed to produce phenol.

6. Make sure that N,N–dimethylaniline is pure on TLC. Since acetic acid has a strong odor, the addition should be carried out under the hood.

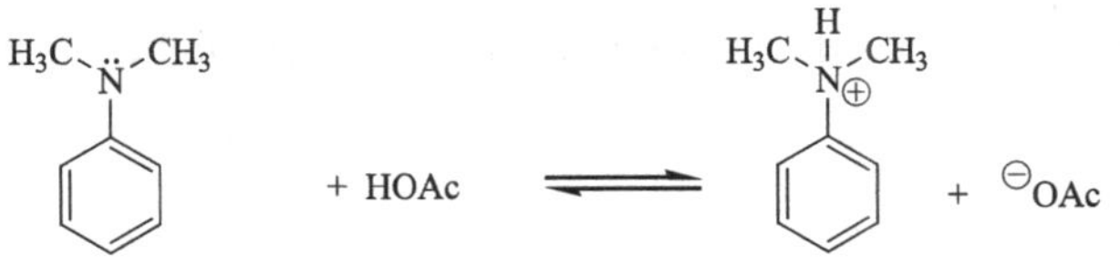

7. Because of its acidic condition, methyl orange appears light red.

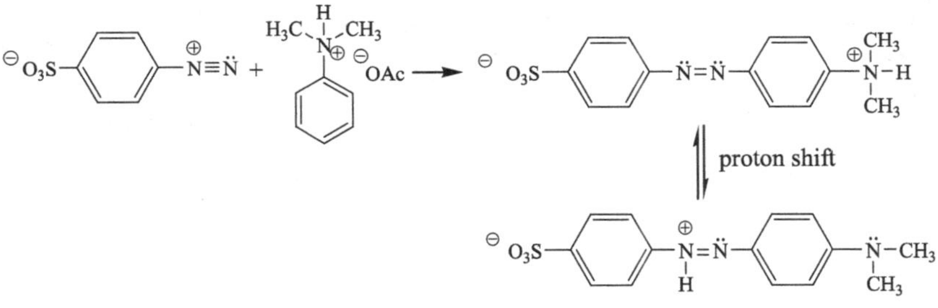

8. Add about 2.0~3.0 mL of 10 % sodium hydroxide solution and check with pH paper. Repeat the process until the pH becomes 8~9.

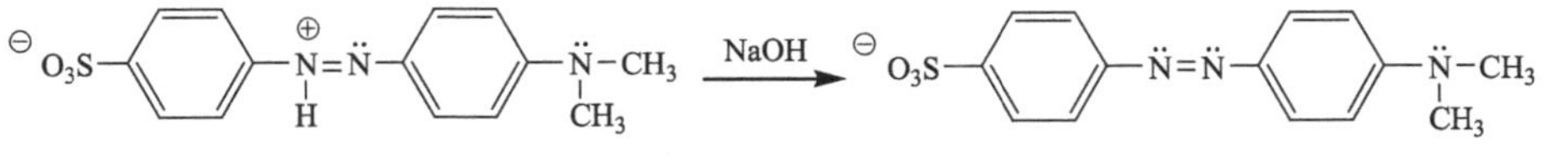

9. The solubility of organic substances in water is markedly reduced by the presence of dissolved inorganic salts, especially sodium chloride. (**Salting–out effect,** refer to **1.4 Extraction**)
By saturating water with sodium chloride, the yield can be increased.

10. Place the reaction mixture in an ice bath to increase the yield. Wash it with the saturated sodium chloride solution since methyl orange shows a high solubility in water.

11. It takes a long time to dry the product completely in this reaction. Instead of calculating the yield, check out the color change according to the pH.

12. Methyl orange is an **acid–base indicator**. It is yellow, and λ_{max} is 460 nm in a dilute solution with a pH higher than 4.4. The protonated form is 520 nm and appears red.

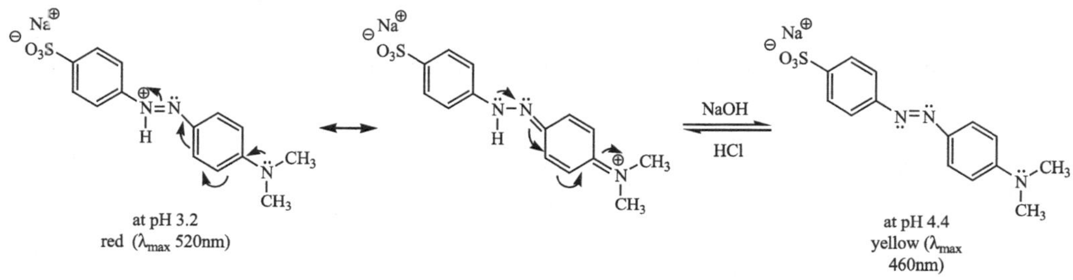

Chemical equation

Experiment 2

Reagents

Name	Molecular formula	Structural formula	M. W.	m.p. (℃)	b.p. (℃)	Density (g/mL)
Sodium carbonate						
Sulfanilic acid						
Sodium nitrite						
2-Naphthol						
Orange II (Product)						

Apparatus

⇨ magnetic stirrer, magnetic spin–bar, lab jack, ice bath

Glassware

⇨ 250 mL Erlenmeyer flask, 50 mL Erlenmeyer flask, 100 mL graduated cylinder, 250 mL beaker, Büchner funnel, filter flask, test tube

Experimental procedure

Because this reaction deals with the salt type of reaction intermediates, it is difficult to monitor the reaction with TLC. Therefore, observing the color change of the reaction mixture is required for monitoring the reaction.

<Making diazonium salt>

1. In a 250 mL Erlenmeyer flask **A**, place 100.0 mL of distilled water.
   Add 275mg (2.6 mmol, 0.46 eq) of sodium carbonate and 1.00 g (5.8 mmol, 1.0 eq) of sulfanilic acid and heat the reaction mixture with stirring until it dissolves completely. [Note 1]
2. When the reaction mixture dissolves completely, cool it down to room temperature and then place it in the ice bath. [Note 2]
3. Add 375mg (5.4 mmol, 0.96 eq) of sodium nitrite. [Note 3]
   (Observation 1: change in temperature)
4. Add 1.0 mL (1.20 g, 11.5 mmol, 2.1 eq) of concentrated hydrochloric acid to 10.0g of ice in a 50 mL Erlenmeyer flask **B.**
   Add ice–cold hydrochloric acid in the flask **B** to the flask **A** slowly with a disposable pipette. [Note 4]
5. Stir the resulting mixture thoroughly in the ice bath until the diazonium salt is formed. [Note 5] (Observation 2: change in color)

<Synthesis of Orange II>

6. In a test tube, mix 0.80 g (5.6 mmol, 1.0 eq) of 2–naphthol and 5.0 mL (15.0 mmol, 2.7 eq) of 3.0 M sodium hydroxide solution. [Note 6]
   Add the mixture to the flask A from the procedure 5.
   (Observation 3: change in color)
7. Heat the resulting reaction mixture with stirring for 20 minutes. [Note 7]

Date of experiment : **Room temperature:**

Observation

Observation 1 :

Observation 2 :

Observation 3 :

<Work–up>

8. Cool down the reaction mixture to room temperature.
   Add 10.0 g of sodium chloride and then place the reaction mixture in the ice bath. [Note 8] (Observation 4: state of solution)

9. When no more crystal is precipitated, filter it by the appropriate suction filtration procedure that is dealt in **1.5 Recrystallization** and wash it with 20.0 mL of brine. [Note 9]
   (Observation5: weight of filter paper)

10. After the product is sufficiently dried in air, weigh the product and calculate the percent yield. [Note 10, 11]
    (Observation 6: weight, color, and percent yield)

Observation 4 :

Observation 5 :

Observation 6 :

- Theoretical weight of product ____________ g
- Practical weight of product ____________ g
- Percent yield ____________ %

Checked by ______________________

Note

1. Gently heat the reaction mixture just enough to dissolve it completely.

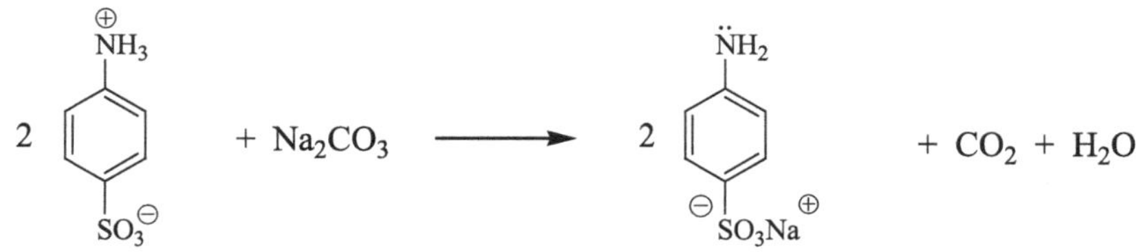

2. While dissolving it, prepare an ice bath in advance.

3. Adjust the amount of sodium nitrite so that the temperature of the reaction mixture does not exceed 10℃.

4. If the temperature rises rapidly, add ice to the reaction mixture to control the temperature.

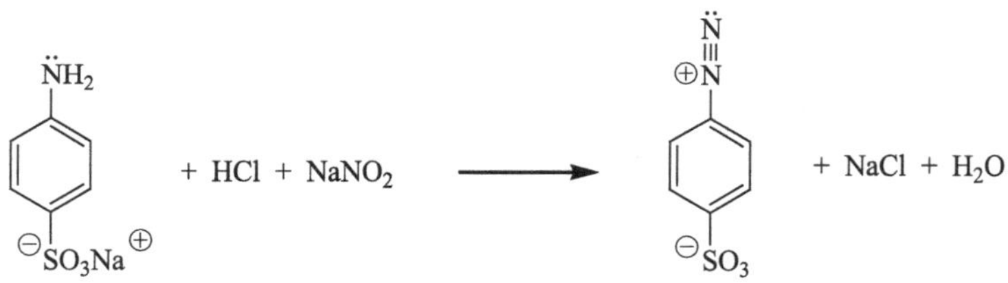

The mechanism for diazonium salt formation is discussed in the previous section **Note 4** on p 226.

5. The color of the reaction mixture will turn to yellow due to the formation of a diazonium salt. The diazonium salt should be kept below 10℃. If the temperature is higher, it will be hydrolyzed to produce phenol.

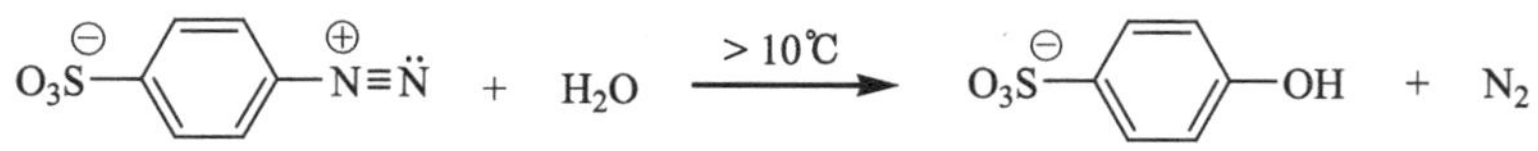

6. Use sodium hydroxide in order to completely dissolve 2–naphtol.

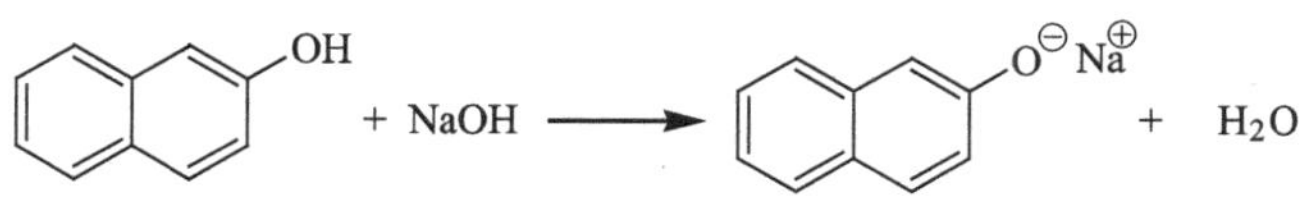

7. Because of its basic condition, Orange II appears light red.

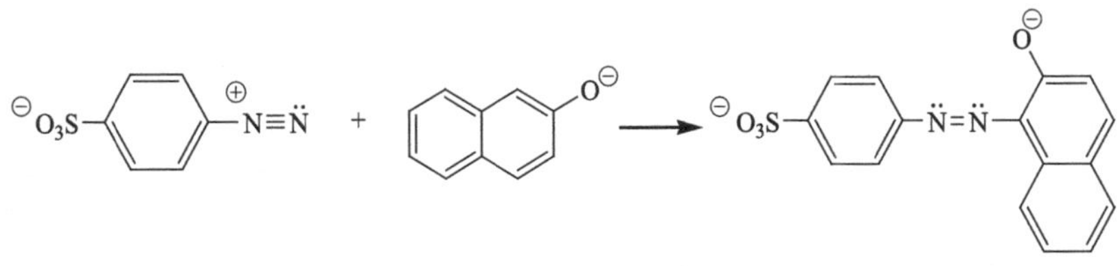

8. The solubility of organic substances in water is markedly reduced by the presence of dissolved inorganic salts especially sodium chloride. (**Salting–out effect,** refer to **1.4 Extraction**)
   By saturating water with sodium chloride, the yield can be increased.

9. Place it in an ice bath to increase the yield. Wash it with the saturated sodium chloride since methyl orange shows a high solubility in water.

10. It takes a long time to dry the product completely in this reaction. Instead of calculating percent yield, check out the color change according to the pH.

11. Orange II is an **acid–base indicator**. It is red at pH 10.2~11.8 and orange at pH 7.4~8.6 in a dilute solution.

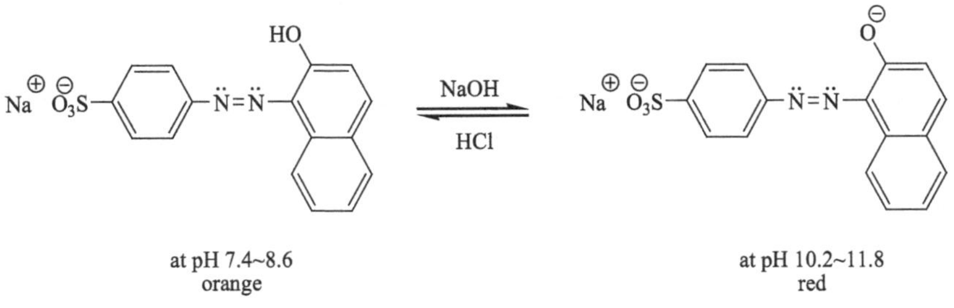

Keywords for Dyes

Chemical equation for Synthesis of Methyl Orange and Orange II

Reaction mechanism for Synthesis of Methyl Orange and Orange II

3.2 Synthesis of Lactam using Beckmann Rearrangement

Keywords : ε-caprolactam, β-lactam, Nylon 6, Nylon 66, oxime, steric hindrance

The **lactam** is a general name for hetero cyclic compounds which contain an amide groups (–CONH–) in a molecule. It is classified according to the size of ring. For example: 4–membered (β–lactam), 5–membered (γ–lactam), 6–membered (δ–lactam), 7–membered ring (ε–lactam) and so on.

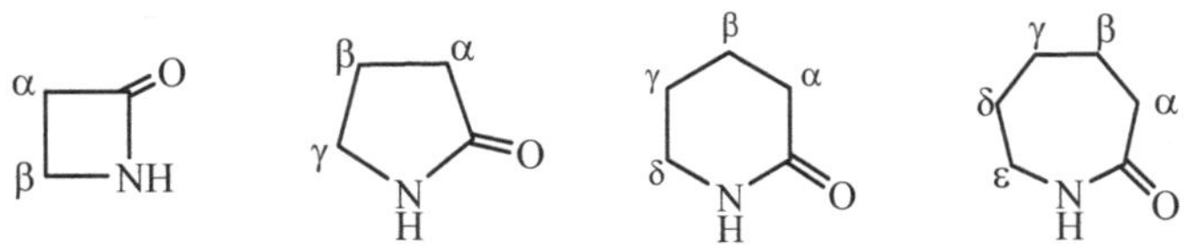

β–lactam γ–lactam δ–lactam ε–lactam

Penicillins and cephalosporins have a β–lactam ring with a thiazolidine ring or a dihydrothiazine ring. They are called β–lactam **antibiotics**, which are major drugs for the treatment of infectious diseases caused by bacteria. The first penicillin to be introduced for therapeutic purposes was benzyl penicillin (penicillin G : $PhCH_2CONH$– at C–6). New β antibiotics have been extensively developed in order to overcome the resistance caused by bacteria. For example, **monocyclic β–lactams** which rarely occur in nature are under active investigation.

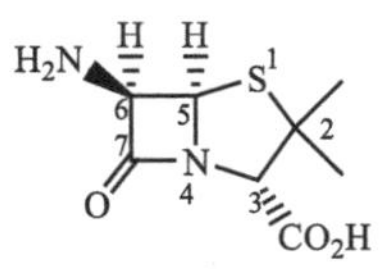

6-APA
6-aminopenicillanic acid

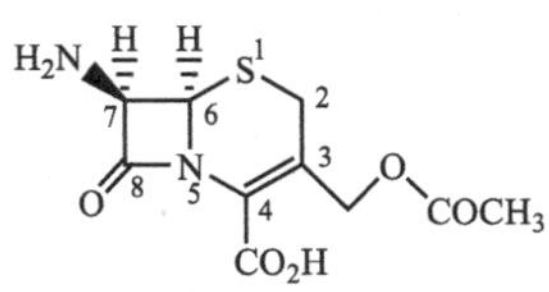

7-ACA
7-aminocephalosporanic acid

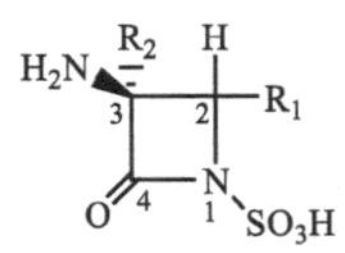

3-AMA
3-aminomonobactam acid

About 90 % of **ε–caprolactam** goes to the manufacture nylon 6 fibers that are used for making clothes, fishing nets and carpets and the remaining 10 % goes to manufacturing plastic for machinery accessories and other engineering plastic. The whole consumption rate is increased by 2~3 % each year.

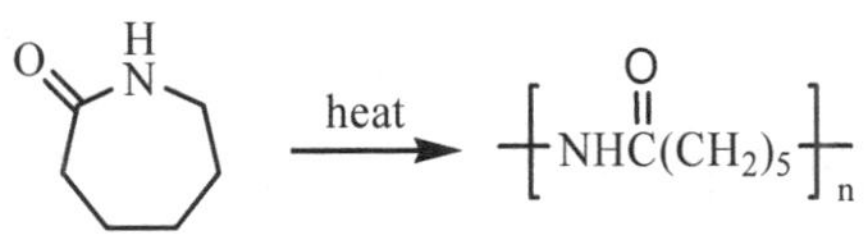

ε–Caprolactam Nylon 6

Nylon 66 is similar to nylon 6 in durability, weight, and chemical resistance, but different in melting point. (Nylon 66 : 250℃, Nylon 6 : 210℃)

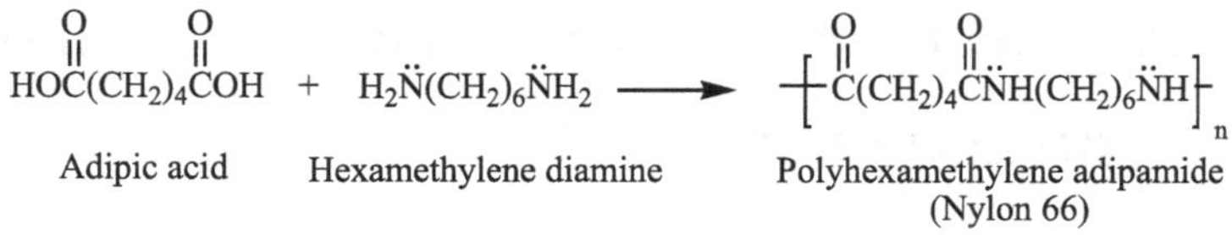

The reaction mechanism for the Beckmann rearrangement starts with protonation or activation of the hydroxyl group. After the hydroxyl group of oxime is converted into a good leaving group, the group at the *trans* position to the resulting leaving group migrates to nitrogen to generate an **electron–deficient intermediate**. It reacts with water to produce an amide or is decomposed to produce a nitrile.

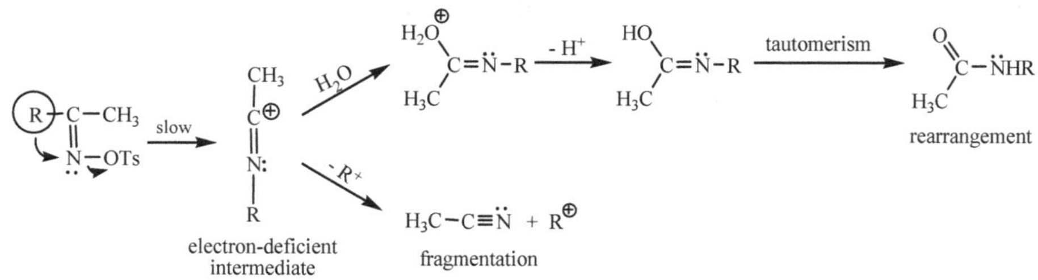

When the quaternary carbon is next to oxime, an abnormal Beckmann fragmentation (or secondary Beckmann rearrangement) occurs competitively. The following examples of abnormal Beckmann fragmentations have been reported.

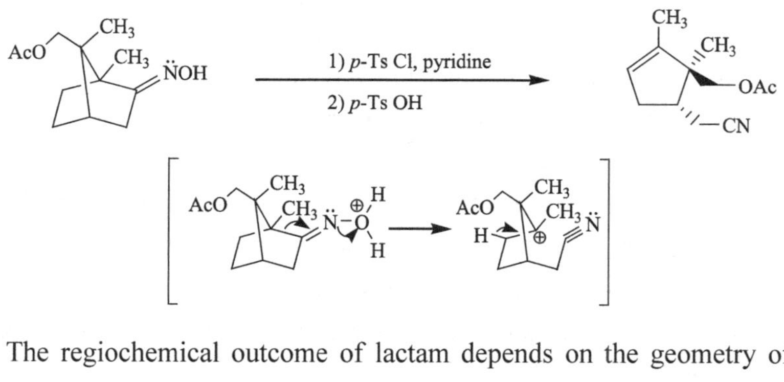

The regiochemical outcome of lactam depends on the geometry of the corresponding oxime.

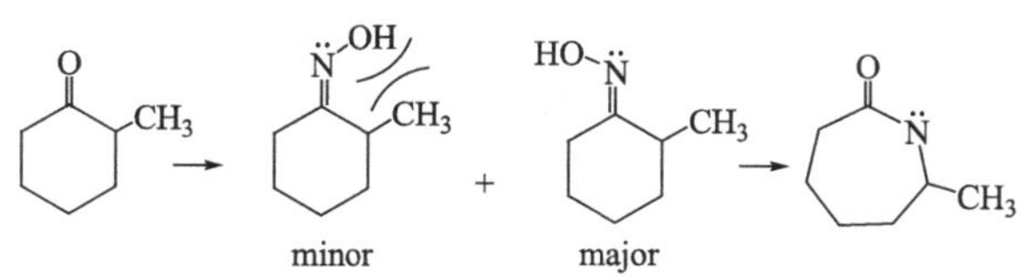

In this experiment, ε–caprolactam is synthesized by Beckmann rearrangement.

Chemical equation

Experiment

Reagents

Name	Molecular formula	Structural formula	M. W.	m.p. (℃)	b.p. (℃)	Density (g/mL)
Cyclohexanone						
Hydroxylamine hydrochloride						
Sodium acetate						
Cyclohexanone oxime						
ε-Caprolactam (product)						

Apparatus

⇨ magnetic stirrer, magnetic spin–bar, lab jack, UV lamp, vacuum pump, evaporator, ice bath, melting point apparatus

Glassware

⇨ TLC chamber, 100 mL one–neck round bottom flask, 50 mL Erlenmeyer flask, 100 mL Erlenmeyer flask, 250 mL Erlenmeyer flask, 100 mL graduated cylinder, 100 mL separatory funnel, conical funnel, watch glass, Büchner funnel, filter flask

Experimental procedure

<Making oxime>

1. In a 50 mL Erlenmeyer flask **A**, place 5.00 g (72.0 mmol, 1.5 eq) of hydroxylamine hydrochloride and 20.0 mL of water.
   (Observation 1: pH)

2. Add 7.50 g (91.4 mmol, 1.9 eq) of sodium acetate in the flask **A**. [Note 1] (Observation 2: pH)

3. Add 5.0 mL (4.74 g, 48.2 mmol, 1.0 eq) of cyclohexanone to the reaction mixture in the flask **A**.
   To monitor the reaction, prepare 3 or 4 TLC plates and spot them with the starting material. [Note 2, 3] (Observation 3: state of solution)

4. Check the reaction with TLC every 10 minutes.
   <the developing solvent is ethyl acetate: hexane = 1:2> [Note 4] (Observation 4: TLC)

5. When cyclohexanone is completely disappeared on TLC, continue stirring for another 5 minutes.
   Filter it by the appropriate suction filtration procedure that is dealt in **1.5 Recrystallization**.
   Remove water as much as possible from the collected solid by squeezing it between sheets of filter paper.
   About 6.0 g of cyclohexanone oxime can be obtained.
   Use it promptly in the next reaction – the Beckmann rearrangement experiment.

Date of experiment : **Room temperature:**

Observation

Observation 1 :

Observation 2 :

Observation 3 :

Observation 4 :

<Beckmann rearrangement>

6. In a 100 mL Erlenmeyer flask **B** in an ice bath, place 16.5 g (292 mmol) of potassium hydroxide in 50.0 mL of water in order to prepare 33 % potassium hydroxide solution. (5.9 M KOH solution)

7. Place 2.0 ml of water in a 50 mL beaker **C** in an ice bath and add 8.0 mL of concentrated sulfuric acid slowly in order to prepare 85 % sulfuric acid solution. (14.6M)

8. Place 3.00 g (26.5 mmol, 1.0 eq) of the crude cyclohexanone oxime from the procedure 5 in a 50 mL Erlenmeyer flask **D**.

9. **<Caution> This process must be supervised by the laboratory instructor and must be carried out under the hood.**

   Add 4.0 mL of 85 % sulfuric acid solution in the beaker B to the flask **D**.
   When the solid dissolves, heat up the reaction mixture on the hot plate until an abrupt gas eruption is observed. (Observation 5: state of solution)

   ➡ when the solution turns dark brown with bubbles, water vapor begines to condense around the mouth of Erlenmeyer flask. And then it radically boils.

   After observing the gas eruption, heat the reaction mixture for another 30 minutes with stirring and then cool down to room temperature. [Note 5]

10. Place the flask **D** in an ice bath.
    Add 10.0 mL of water to dilute the reaction mixture.

11. Slowly add 33 % potassium hydroxide solution in the flask **B** to the flasks **D** in an ice bath until the pH of the reaction mixture becomes 9~10. [Note 6] (Observation 6: pH)

Observation 5 :

Observation 6 :

<Work–up>

12. Remove the potassium sulfate salt by the appropriate suction filtration procedure that is dealt in **1.5 Recrystallization**. Rinse the reaction flask **D** twice with 5.0 mL of methylene chloride.

13. Transfer the filtrate into a separatory funnel using a conical funnel.

Add 20.0 mL of methylene chloride. Extract and then transfer the organic layer to another separatory funnel.
Extract the aqueous layer again twice with 20.0 mL of methylene chloride each time.
Check the aqueous layer with TLC, and if the product remains, extract by adding another 20.0 mL of methylene chloride.
<the developing solvent is ethyl acetate> (Observation 7 : TLC)

16. The organic layers from the procedure 12 and 13 are combined, washed with brine and then drawn off carefully through the stopcock into a clean Erlenmeyer flask **E**. Rinse the stem with 10.0 mL of methylene chloride. Add a small amount of anhydrous magnesium sulfate ($MgSO_4$) to the organic layer in the flask **E** until the solution becomes clear and the solid $MgSO_4$ moves freely in the solution.

17. The resulting organic layer is filtered according to the appropriate suction filtration procedure, which is dealt in **1.5 Recrystallization**.

18. Transfer the resulting solution (**filtrate**) to a pre–weighed 100 mL one–neck round bottom flask (**tared flask**). Evaporate methylene chloride by using a rotary evaporator. When the solvent is almost evaporated, ε–caprolactam will remain at the bottom of the flask.

19. Check the purity of the product using TLC and find percent yield. [Note 7,8] (Observation 8: TLC, m.p. and percent yield)

Observation 7 :

Observation 8 :

- Theoretical weight of product ____________ g
- Practical weight of product ____________ g
- Percent yield ____________ %

Checked by ______________________

Note

1. The optimal pH of the reaction medium for this reaction is between pH 4 to 5. When the reaction medium is too acidic, the lone–pair electrons on the nitrogen atom of H_2NOH readily attack the proton to form a quaternary ammonium salt. As a result, the nitrogen loses its nucleophilicity and the initial addition step (I ⟶ II) does not occur properly.

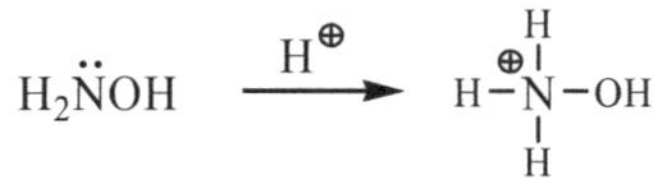

2. On the other hand, if the reaction medium is too basic, the irreversible dehydration step (IV ⟶ V) does not occur since the key intermediate IV (**oxonium**) from III is not formed.

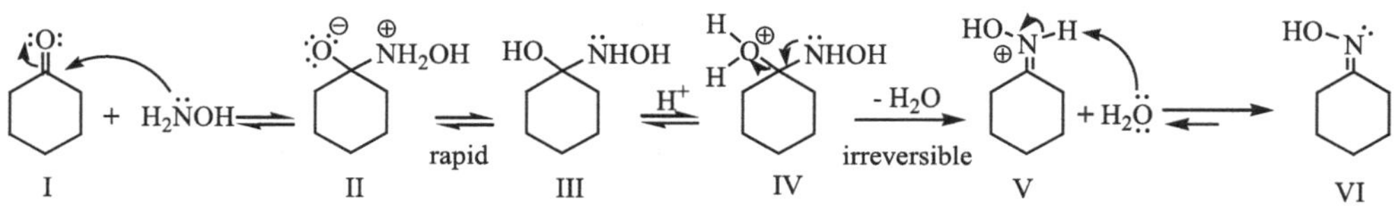

3. When the acetate is added to the reaction medium, the optimized pH condition pH 4~5 is achieved. Sodium acetate is a weak base and removes the proton from the nitrogen atom of hydroxylamine hydrochloride. The resulting hydroxylamine is ready to attack the carbonyl group of the cyclohexanone.

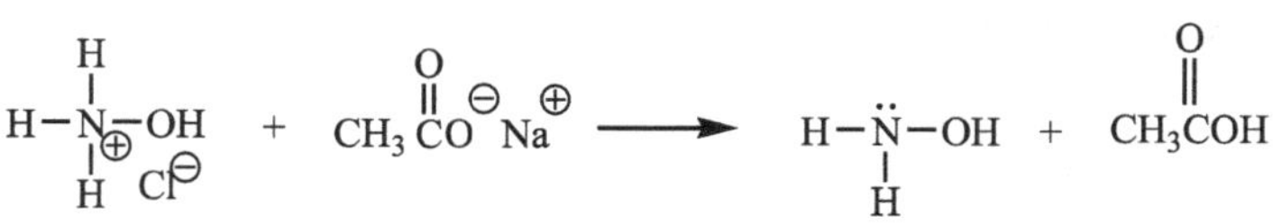

4. The reaction proceeds rapidly. (10~20 minutes)
Since cyclohexanone, cyclohexanone oxime and ε–caprolactam are not active under the UV light, use potassium permanganate ($KMnO_4$) dyeing reagent and check the stained spots.

5. The violently evolving gas is the water vapor. Removal of water is the **driving force** for the Beckmann rearrangement reaction.

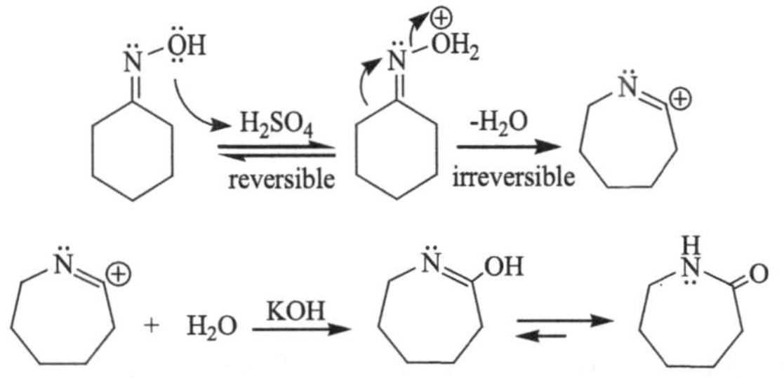

6. The mol ratio of 85 % sulfuric acid (H_2SO_4) and 33 % potassium hydroxide solution (KOH) is 14.6 M: 5.9 M = 2.5 : 1. In theory, 5 times more potassium hydroxide solution is required to neutralize the reaction mixture. About 20.0 mL of 33 % potassium hydroxide solution is needed.

7. Observing m.p. or NMR spectroscopy can identify ε–Caprolactam. The percent yield is influenced by the weight of water that oxime contains. The drier the oxime, the higher the percent yield. The percent yield is 50~80 % based on the cyclohexanone.

    Cyclohexanone oxime
    ^{1}H NMR ($CDCl_3$) : δ1.63 (6H), 2.20 (2H), 2.49 (2H), 7.78 (1H)
    ε–Caprolactam : m.p. 65~67℃ (lit. value : 70~72℃)
    ^{1}H NMR ($CDCl_3$) : δ1.71 (6H), 2.47 (2H), 3.21 (2H), 6.06 (1H)

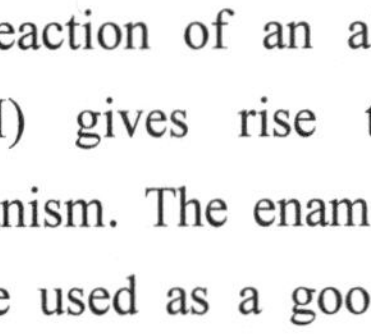

<Chemical shift data of ε–Caprolactam>

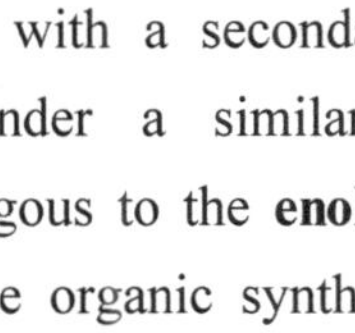

<Predict chemical shift data using Chem Draw>

8. The reaction of an aldehyde or ketone with a secondary amine (R_2NH) gives rise to an **enamine** under a similar reaction mechanism. The enamine, which is analogous to the **enolate anion,** can be used as a good **nucleophile** in the organic synthesis.

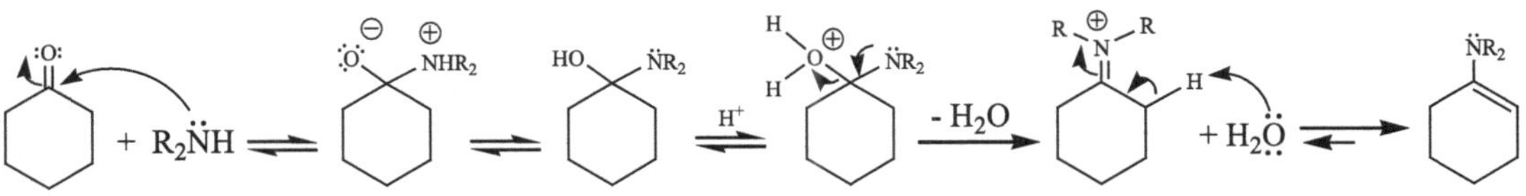

Keywords for Lactams

Chemical equation for Synthesis of ε–Caprolactam

Reaction mechanism for Synthesis of ε–Caprolactam

3.3 Nazarov Reaction

Keyword : pericyclic reaction, conrotatory, disrotatory, enone, 1,2-adduct, 1,4-adduct, Michael acceptor

Pericyclic reactions include the **electrocyclic reaction** (**Nazarov**), the **cycloaddition** (refer to **3.4 Diels-Alder Reaction**), the **sigmatropic rearrangement** (**Cope and Claisen**), and the **chelotropic reaction**.

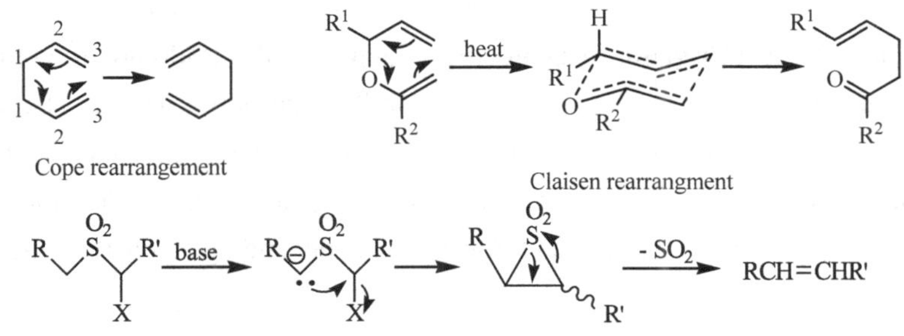

Cope rearrangement

Claisen rearrangment

Chelotropic reaction (Bäcklund-Ramberg reaction)

The **Nazarov cyclization**, is a useful reaction for the synthesis of a **cyclopentenone** skeleton from divinyl, α,β-unsaturated carbonyl compounds. The stereochemistry of the resulting products can be explained by the **molecular orbitals** of the pentadienyl cation based on the **frontier molecular orbital theory**.

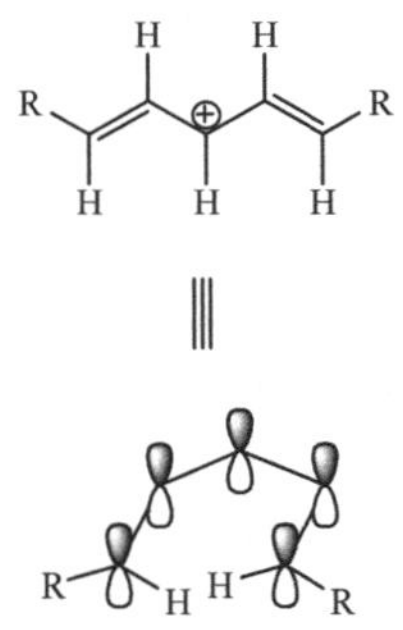

ground state

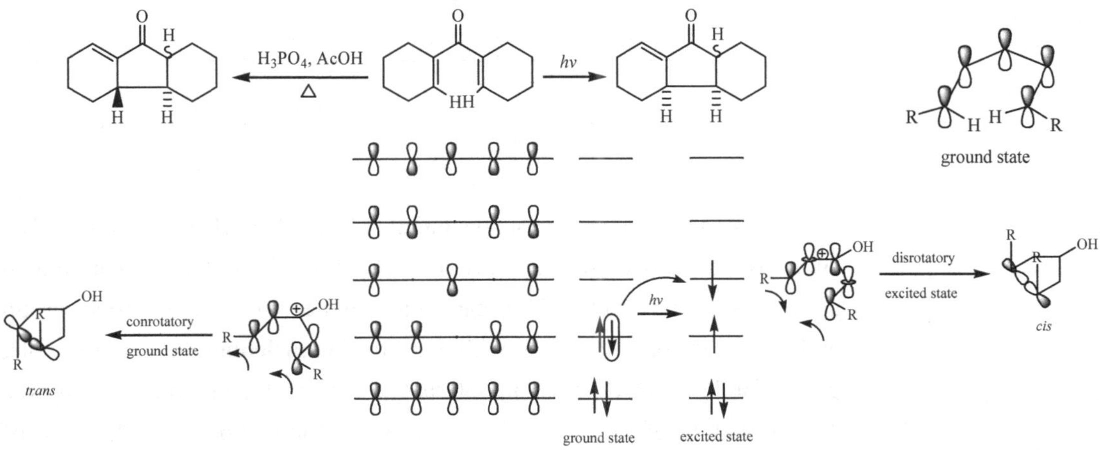

The nucleophilic addition for **enone** is reversible and proceeds to either **1,2-addition** or **1,4-addition**. The **driving force** for addition is the replacement of a π-bond by a σ-bond. It is found that the initial product is a **1,2-adduct** (**kinetic product**). Under the vigorous reaction conditions (higher temperature and longer reaction time), the **1,2–adduct** is converted to a **1,4–adduct (thermodynamic product)**.

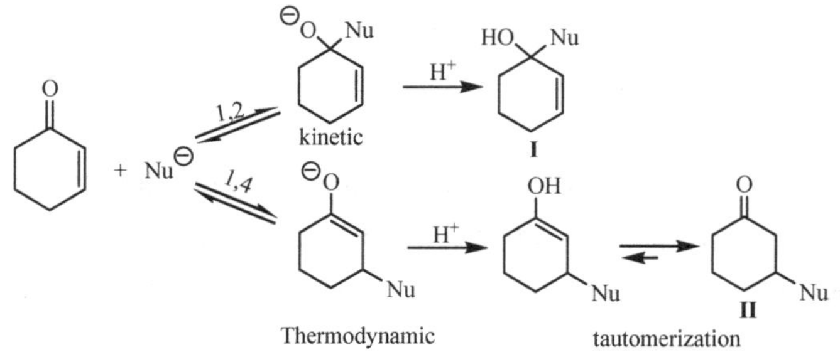

The RMgX (**Grignard reagent**), RLi, and R_2CuLi (**Gilman or cuprate reagent**) are mostly used for the nucleophillic addition reaction. The RLi gives rise to a **1,2–adduct** while the Gilman reagent provides a **1,4–adduct** exclusively. It is reported that the Grignard reagent goes through either 1,2 (I) or 1,4 (II)–addition reaction depending upon the reaction conditions.

$Nu^{\ominus}$	I	II
CH_3Li	95%	5%
$Li(CH_3)_2Cu$	3%	97%

In addition, the 1,4–adduct is predominant at a high reaction temperature since it is a thermodynamic product.

$Nu^{\ominus}$	temp.	I	II
$(CH_3)_2C{=}C(O^{\ominus}Li^{\oplus})OCH_3$	-78°C	88%	5%
	25°C	7%	83%

The 1,4–additional reaction on **enone** is called **Michael reaction.** The enone is also called **Michael acceptor**. The natural substances that contain an enone group show significant bioactivities since the enone group undergoes Michael reaction with various **bionucleophiles**. For example, prostaglandins, which exist in internal organs and body fluids, influence the controlling of the gastric juice secretion, contraction and relaxation of the bronchus muscle with other effects on physiology.

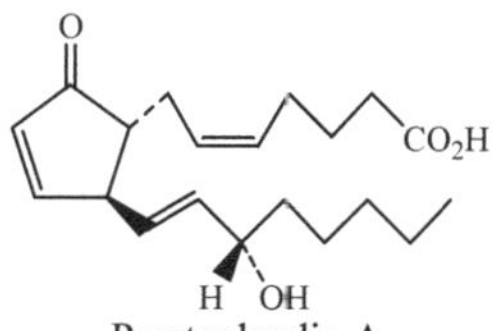
Prostaglandin A_2

In this experiment, we will synthesize 2,3–diphenyl–cyclopenta–2–enone, using *trans,trans*–dibenzylidene–acetone, which was previously prepared in **2.8 Aldol Condensation**.

Chemical equation

Experiment

Reagents

Name	Molecular formula	Structural formula	M. W.	m.p. (℃)	b.p. (℃)	Density (g/mL)
trans,trans-Dibenzylidene-acetone						
Formic acid						
Phosphoric acid						
2,3-Diphenyl-cyclopenta-2-enone (Product)						

Apparatus

⇨ magnetic stirrer, magnetic spin–bar, heating mentle, lab jack, transformer, thermometer

Glassware

⇨ TLC Chamber, 100 mL two–neck round bottom flask, beaker, 100 mL graduated cylinder, 100 mL separatory funnel, filter flask, Büchner funnel, reflux condenser

Experimental procedure

<Reaction>

1. In a 100 mL two–neck round bottom flask **A**, place 1.00 g (4.3 mmol, 1.0 eq) of *trans,trans*–dibenzylideneacetone (refer to **2.8 Aldol Condensation**) and 10.0 mL (12.2 g, 265.0 mmol, 62.1 eq) of formic acid.
   To monitor the reaction, prepare 3 or 4 TLC plates and spot them with the starting material. (Observation 1: actual amount added)
2. In the flask **A**, add 10.0 mL (16.8 g, 150.0 mmol, 35.1 eq) of phosphoric acid slowly.
   Assemble the reflux apparatus on the flask **A**.
   (refer to **1.3 Distillation**)
   Reflux the reaction mixture in the flask A [Note 1]
3. Check the reaction mixture with the pre–spotted TLC plate every one hour. <the developing solvent is ethyl acetate: hexane = 1:2> (Observation 2: TLC)
   <Caution> A type of burn may occur by hot vapor of phosphoric acid. Make sure the reaction mixture is cooled down enough, and wear gloves when taking the sample for TLC from the flask **A**. [Note 2]
4. When *trans,trans*–dibenzylideacetone is completely disappeared on TLC, continue stirring for another 30 minutes.
   Cool down the reaction mixture in the flask A and add 30.0 mL of water through the top of the water–cooled condenser.
   Reflux it for another hour and cool down to room temperature.

Date of experiment : **Room temperature :**

Observation

Observation 1 :

Observation 2 :

<Work up>

5. Disassemble the reflux apparatus.
   Add 20.0 mL of methylene chloride and stir well.

6. Transfer the resulting reaction mixture to the separatory funnel using a conical funnel and shake well. Allow it to stand undisturbed until the layers are completely separated.
   Transfer the organic layer to another separatory funnel.
   Check if the reaction mixture remains in the aqueous solution by TLC. If necessary, extract the aqueous layer again. [Note 3]
   (Observation 3: actual amount of solvent used, TLC).

7. The organic layers are combined, washed with brine and then drawn off carefully through the stopcock into a clean Erlenmeyer flask **B**. Rinse the stem with 10.0 mL of methylene chloride.
   Add 20.0 mL of the saturated sodium bicarbonate in the flask **B** and stir thoroughly. Check pH of the resulting reaction mixture. Add more the saturated sodium bicarbonate until it becomes pH = 7~8. [Note 3]
   (Observation 4: volume of used saturated sodium bicarbonate and pH)

8. Transfer the organic layer in the flask **B** to a separatory funnel **C** using a conical funnel and extract. The resulting organic layer is transferred to a separatory funnel **D,** washed with brine and then drawn off carefully through the stopcock into a clean Erlenmeyer flask **E**. Rinse the stem with 10.0 mL of methylene chloride.
   Add a small amount of anhydrous magnesium sulfate ($MgSO_4$) to the organic layer in an Erlenmeyer flask until the solution becomes clear and the solid $MgSO_4$ moves freely in the solution.

Observation 3 :

Observation 4 :

9. The resulting organic layer is filtered according to the appropriate suction filtration procedure, which is dealt in **1.5 Recrystallization**.

10. Transfer the resulting solution (**filtrate**) to a pre–weighed 100 mL one–neck round bottom flask (**tared flask**). Evaporate methylene chloride by using a rotary evaporator. When the solvent is almost evaporated, 2,3–diphenyl–cyclopenta–2–enone will remain at the bottom of the flask. [Note 4]

10. Check the purity of the obtained product with TLC, and calculate percent yield. (Observation 5: TLC and percent yield) [Note 5,6,7]

Observation 5 :

- Theoretical weight of product __________ g
- Practical weight of product __________ g
- Percent Yield __________ %

Checked by ____________________

Note

1.

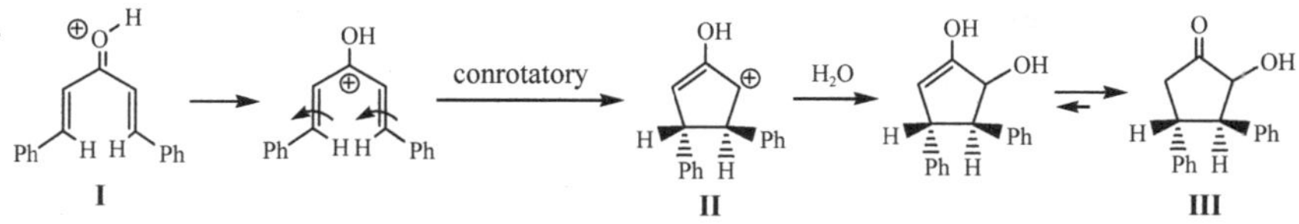

The carbonyl group of **I** (*trans,trans*–dibenzylideneacetone) is activated to an oxonium ion through protonation by the acid catalyst. The electrocyclic reaction under the thermal condition transforms **I** to **II**. (hydroxy allyl cation) The resulting **II** is readily transformed to an **III** (acyloin) by the attack of the hydroxylic solvent (e.g. water).

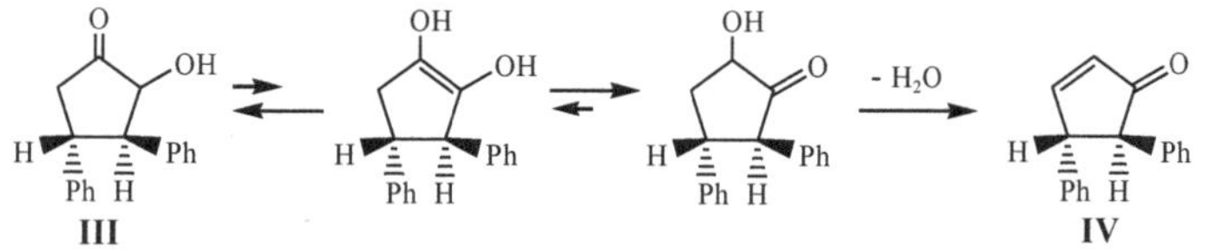

Then the following sequence of **isomerization**, **tautomerization** and **dehydration** gives rise to the key intermediate **IV**.
IV is further converted through a series of tautomerizations to a more stable isomer **V**.

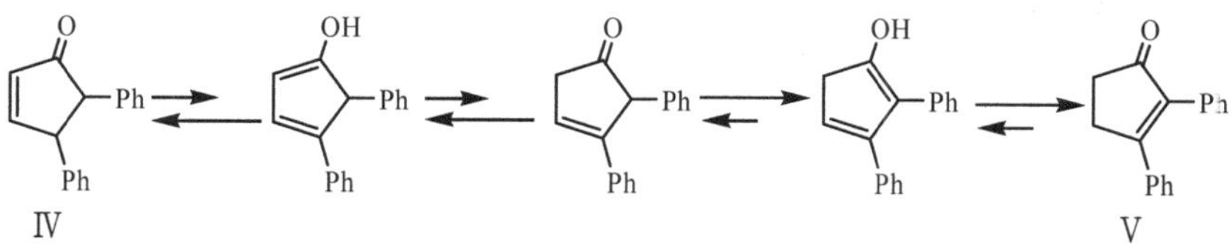

2. When the reaction is checked after two hours, most of the starting substances should have disappeared. Even if a small amount of the starting substances is remained, proceed to the next step.

3. If a mixture remains in the aqueous solution, extract it by using methylene chloride. The organic layer should be washed with water in advance to prevent any unexpected rapid effervescence on adding the saturated sodium bicarbonate solution.

4. After the rotary evaporation, a brown thick mixture is obtained. The percent yield of the crude product is about 90~110 %.

5. The crude product shows 3 spots on TLC under the UV lamp. Among them, the first spot (R_f = 0.75) is the starting material. Between the very small second spot (R_f = 0.64) and the biggest third spot (R_f = 0.5), the third spot is the desired product. However, the first and the second spots do not have much effect on the purity of main product for taking NMR spectrum.

6. You can observe the formation of the main product in the (200MHz) NMR spectrum of the crude product.

2,3–Diphenyl cyclopenta–2–enone
^{1}H NMR ($CDCl_3$) : δ 2.72 (t, *J*=6.5Hz, 2H), 3.07 (t, *J*=6.5Hz, 2H), 7.36 (m, 10H)

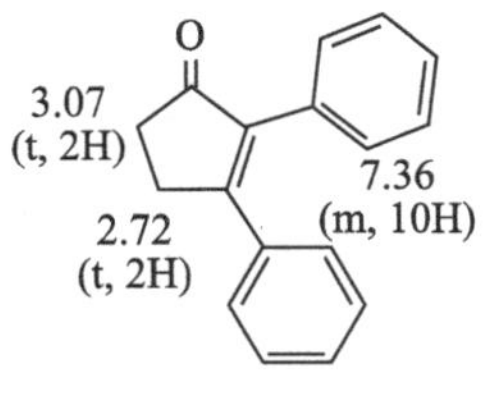

<Chemical shifts of 2,3-Diphenyl cyclopenta-2-enone>

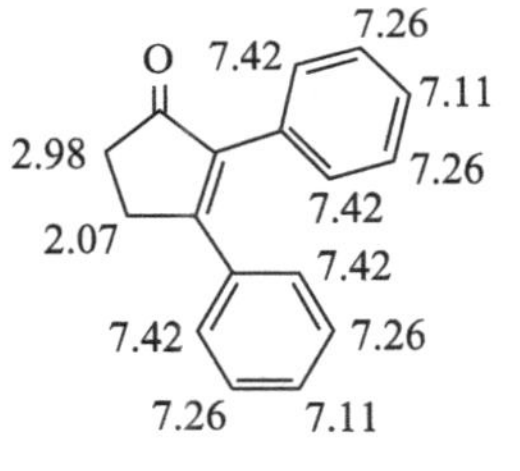

<Prediction of chemical shifts through ChemDraw>

7. When column chromatography is carried out on the crude product, the pure product can be obtained in about 70 %.
The yield obtained by the evaporated product is higher since methylene chloride is not completely removed by the rotary evaporation.

Keywords for Nazarov reaction

Chemical equation for Synthesis of 2,3–Diphenyl cyclopenta–2–enone

Reaction mechanism for Synthesis of 2,3–Diphenyl cyclopenta–2–enone

3.4 Diels-Alder Reaction

Keywords : diene, dienophile, stereospecific, stereoselective, *endo*, *exo*, diastereofacial selectivity, cracking

The **Diels–Alder reaction** is a 4+2 reaction between the conjugated **diene** and the **dienophile**. It produces a 6–membered ring. This is an interesting reaction, which is similar to the resonance contributor of the benzene ring.

Resonance structure of benzene

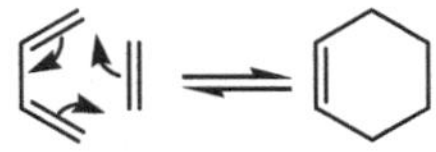

Diels-Alder reaction

The Diels–Alder reaction undergoes a **concerted reaction mechanism** in which new bonds between the diene and the dienophile are formed simultaneously.

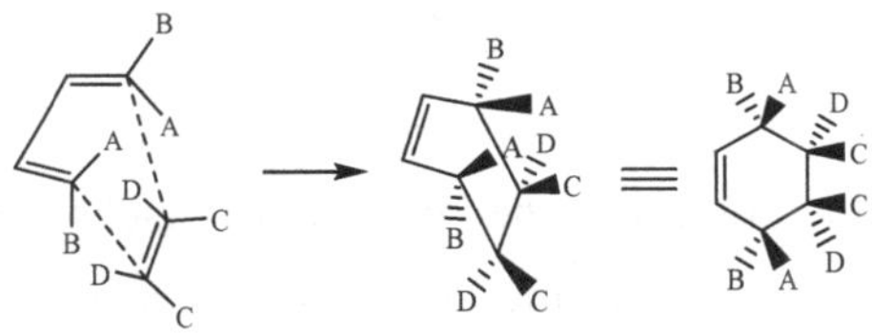

According to the "***cis*** **principle**," the relative stereochemistry of the substituents in both the dienophile and the diene are retained in the adduct.

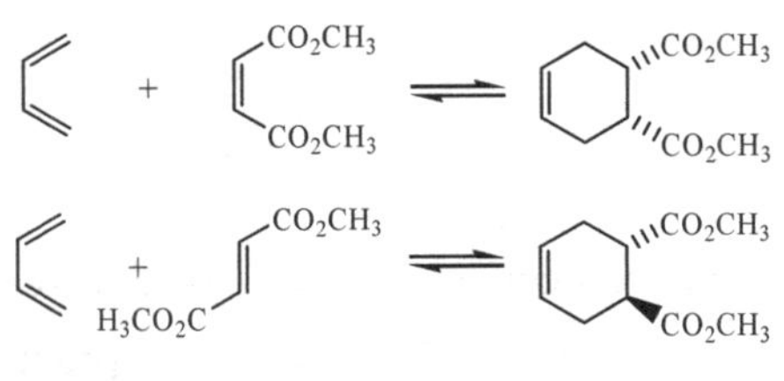

This is a good example of the **stereospecific** reaction.

The smaller the energy separation between **HOMO** (highest occupied molecular orbital) and **LUMO** (lowest unoccuppied molecular orbital), the faster the Diels–Alder reaction. In principle, it is possible to increase the rate by adding the electron donating group (EDG) to the diene and the electron withdrawing group (EWG) to the dienophile since the EDG raises and the EWG lowers the energy both in HOMO and LUMO. ($\Delta E' < \Delta E$)

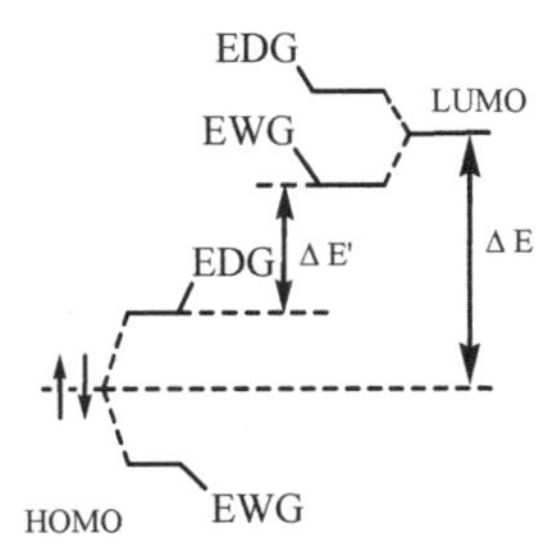

ΔE'< ΔE

The cycloadditon of 1–substituted butadienes with dienophiles such as α,β–unsaturated carbonyl compounds predominantly gives rise to the ***ortho*–adduct**, whereas the cycloaddition of 2–substituted butadienes predominantly leads to the ***para*–adduct**.

This result can be explained by the frontier molecular orbital theory. The **regiochemistry** of the product is largely governed by the **molecular orbital coefficients** at the terminus of the diene and the dienophile.

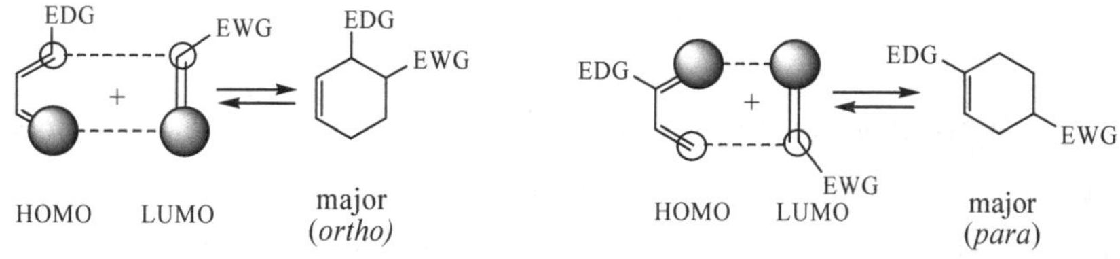

The **Danishefsky diene** is one of the most valuable dienes which fulfill the requirements for an active diene and a good starting material.

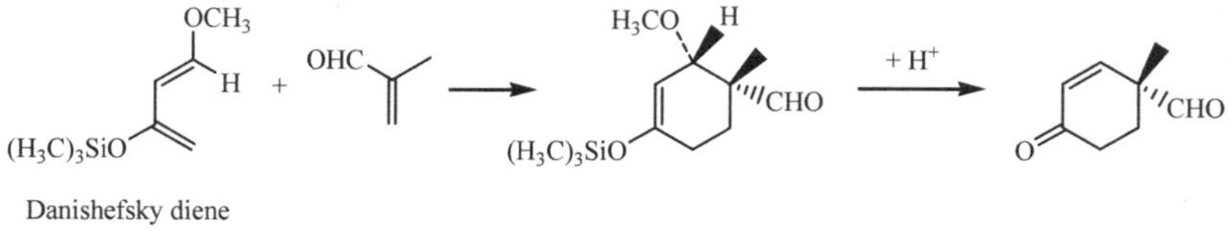

In the Diels–Alder reaction, the dienophile attacks from either above or below of the diene. Under the achiral reaction environment, it gives rise to both **enantiomers** – the **racemic mixture** –.

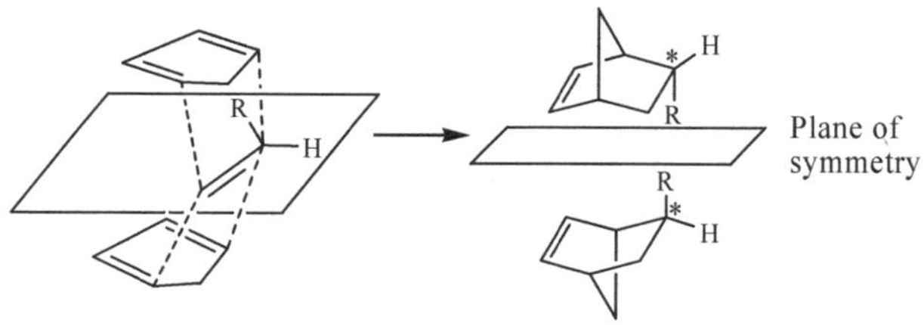

When a chiral environment is applied to the reaction system, it selectively attacks from the preferred direction producing one of the **diastereomers in excess. (d.e.)**

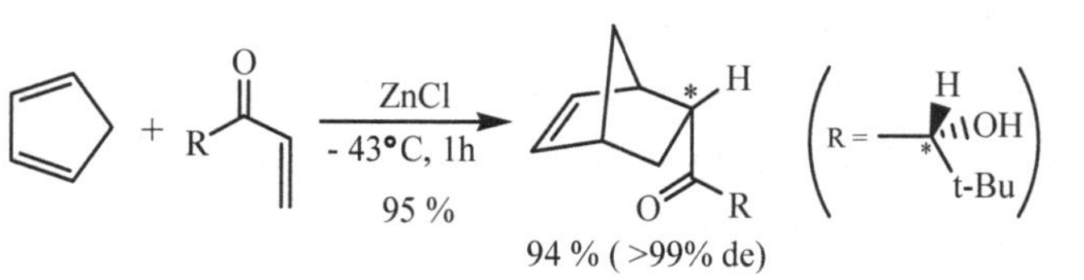

This is a good example of the **stereoselective** reaction based on the **diastereofacial selectivity**.

In this experiment, we obtain cyclopentadiene by cracking dicyclopentadiene and then proceed to the Diels–Alder reaction between cyclopentadiene and maleic anhydride.

Chemical equation

Experiment

Reagents

Name	Molecular formula	Structural formula	M. W.	m.p. (℃)	b.p. (℃)	Density (g/mL)
Dicyclopentadiene						
Maleic anhydride						
cis–Norbornene–5,6–*endo* dicarboxylic anhydride (Product)						

Apparatus

⇨ magnetic stirrer, magnetic spin–bar, ice bath, lab jack, transformer, heating mentle, syringe, melting point apparatus

Glassware

⇨ TLC chamber, 50 mL one-neck round bottom flask, 100 mL two-neck round bottom flask, beaker, Büchner funnel, filter flask, distillation set, reflux condenser

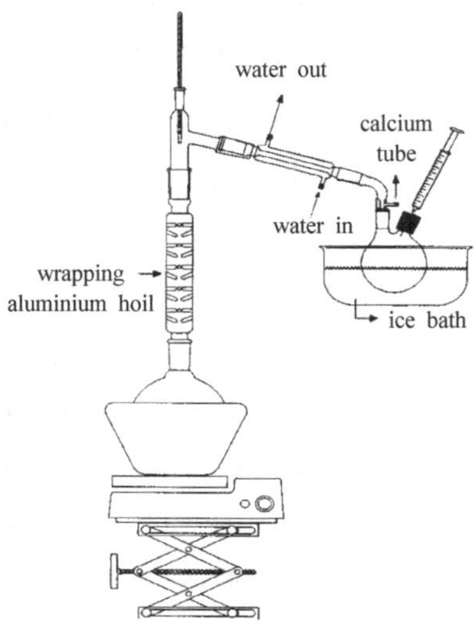

<Fractional distillation apparatus for cracking dicyclopentadiene>

Experimental procedure

– This experiment must be carried out under the hood. –

<Cracking>

1. Place 20.0 mL (19.7g, 11.6 mmol) of dicyclopentadiene and a spin–bar in a 50mL one–neck round bottom flask **A**.
   Assemble the distillation apparatus on the flask **A**. (refer to **the setting diagram in the previous page**)
   Before starting distillation, make sure that the receiving two–neck round bottom flask **B** is placed in an ice bath. Keep it in the ice bath throughout the experiment. [Note 1, 2]
2. Cyclopentadiene is collected at the temperature range of 40~43℃. After cracking, about 10 mL (7.91 g, 12.0 mmol) of cyclopentadiene can be obtained. Keep it in an ice bath. [Note 3 or 4]

<Reaction>

3. In a 100mL two–neck round bottom flask **C**, place 4.00 g (40.8 mmol, 1.0 eq) of maleic anhydride in 10.0 mL of ethyl acetate and stir well. After maleic anhydride dissolves completely, add 10.0 mL of hexane and place the flask **C** in an ice bath.
   To monitor the reaction, prepare 3 or 4 TLC plates and spot them with the starting material. [Note 5]
4. Draw out 4.0 mL (3.20 g, 48.9 mmol, 1.2 eq) of cyclopentadiene from the flask **B** using a syringe, and then slowly add it to the flask **C**.
   [Note 6] (Observation 1 : formation of a solid)
5. Check the reaction mixture every 20 minutes with the pre–spotted TLC plate. <the developing solvent is acetone> [Note 7]
   As maleic anhydride and the product do not appear under UV light, use a dyeing reagent (potassium permanganate, $KMnO_4$) and observe the stained spots. [Note 8]
6. When maleic anhydride is completely disappeared on TLC, continue stirring for another 20 minutes. (Observation 2 : TLC)

Date of experiment : **Room temperature :**

Observation

Observation 1 :

Observation 2 :

<Work–up>

7. Filter the reaction mixture in the flask **C** by the appropriate suction filtration procedure that is dealt in **1.5 Recrystallization**. The crystal is stick–shaped and white–colored.
8. Most of the solvent can be removed by squeezing the solid between sheets of filter paper before allowing them to dry in the air. Dry the obtained solid before the next laboratory period.
9. Calculate the percent yield and measure the melting point. [Note 9,10] (Observation 3: weight, percent yield and melting point)

Observation 3 :

- Theoretical weight of product ____________ g
- Practical weight of product ____________ g
- Percent yield ____________ %

Checked by ______________________

Note

1. When setting up for **cracking**, the condenser of the distillation set should be longer than normal. In order to prevent the heat loss, the condenser should be covered with aluminum foil and/or glass wool. Use a syringe for removing cracked cyclopentadiene, which should be used immediately. The extra cracked cyclopentadiene should be kept in an ice bath.

2. The basic definition of the **Gibbs–Helmholtz free energy equation** is

$$\triangle G = \triangle H - T\triangle S$$

① if G is negative, the reaction is spontaneous.
② if G is positive, the reaction will not occur spontaneously Instead, the reverse reaction will occur.
③ if G is 0, the system is at equilibrium

The Diels–Alder reaction is a reversible reaction

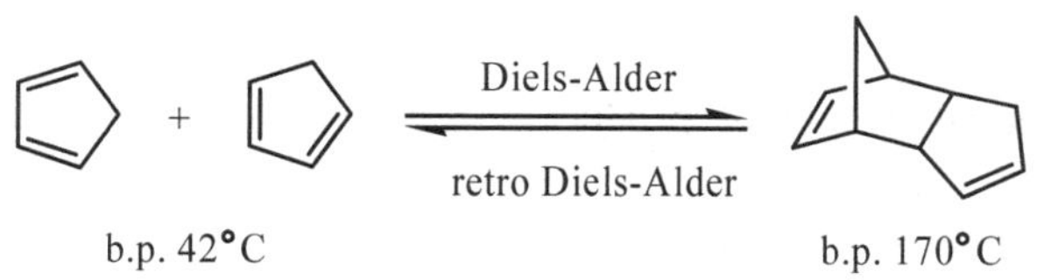

The **enthalpy** for the forward reaction is negative (exothermic) since two σ bonds (376 kJ/mol) are formed at the expense of two π bonds (235 kJ/mol). It contributes to the negative value of the free energy $\triangle G$.

Meanwhile, the **entropy** for this transformation is negative since two molecules are combined to form a single molecule (loss of freedom). $-T\triangle S$ term contributes to the positive value of free energy $\triangle G$. If Diels–Alder reaction is at equilibrium, $\triangle G$ is 0

$$\triangle G = \triangle H - T\triangle S = 0$$

$$\triangle H = T_{eq}\, \triangle S$$

$$T_{eq} = \frac{\triangle H}{\triangle S}$$

When the reaction temperature T is lower than T_{eq}, the enthalpy term is dominant over the entropy term resulting in a negative $\triangle G$.

The forward reaction is spontaneous.

On the contrary, the entropy term is dominant over the enthalpy term when the reaction temperature is higher than T_{eq}. The resulting △G is positive and the reverse reaction proceeds.

Take a simple numerical example for a better understanding:
When a system is △H = –280 kJ and △S = –0.75 kJ/K in equilibrium at T= 373 K,
If T = 250 K
△G = –280 kJ – (250 K) × (–0.75 kJ/K) = –93.5 kJ
The forward reaction is spontaneous.
If T = 400 K
△G = –280 kJ – (400 K) × (–0.75 kJ/K) = +20 kJ
The reverse reaction proceeds alternatively.

3. The cracking reaction of dicyclopentadiene is a good example of the **retro Diels–Alder reaction**. One of the retro Diels–Alder reactions is the following:

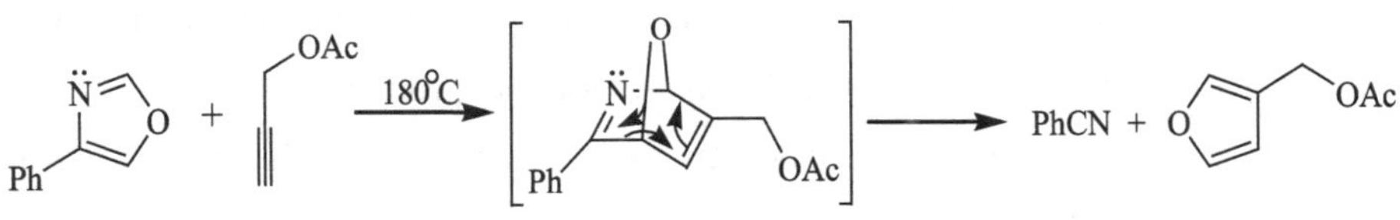

4. Calculate the amount of cyclopentadiene as 70~75 % of the amount of dicyclopentadiene, and add the appropriate amount of dicyclopentadiene in a 50 mL one–neck round bottom flask.

5. If a solid forms on adding hexane, add 5.0 mL of ethyl acetate and dissolve it completely. The reason for adding hexane is to direct the reaction forward since the product will not dissolve in hexane and precipitate out from the reaction mixture. (**Le Châtelier's principle**)

6. The addition of cyclopentadiene is an exothermic reaction. Therefore, carry out the reaction after cooling cyclopentadiene sufficiently in an ice bath.

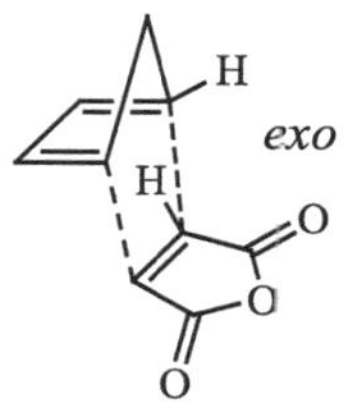

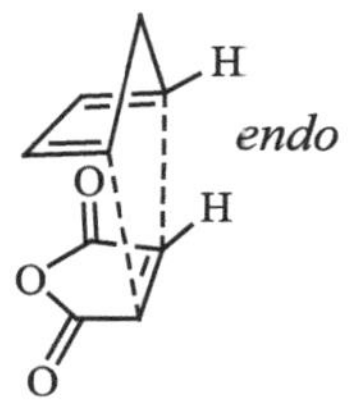

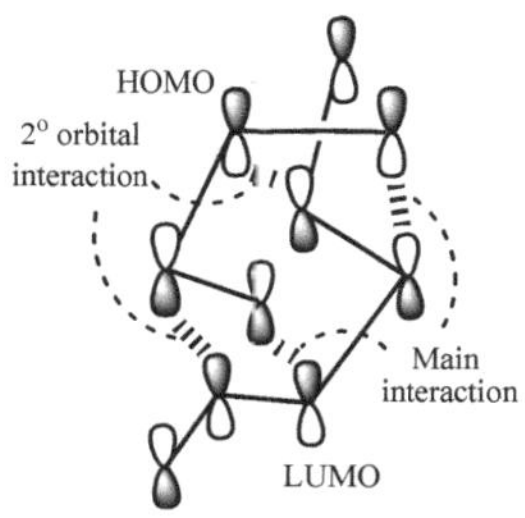

endo transition state

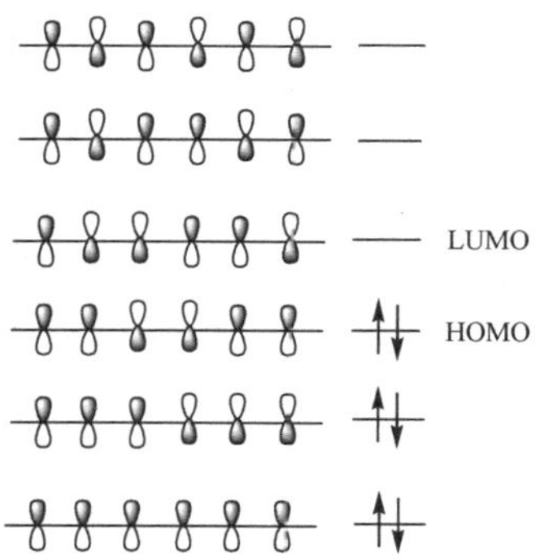

7. The maleic anhydride and the cyclopentadiene give rise to the ***endo*–adduct** as the major product even though the ***exo*–adduct** is thermodynamically more stable.

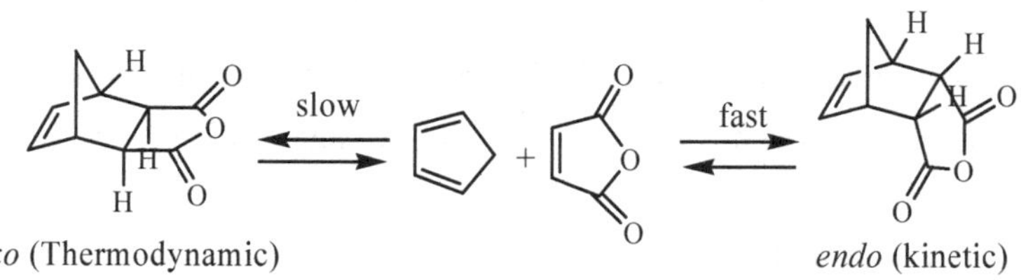

The transition state includes the **secondary orbital overlap interaction** between the diene and the dienophile where carbons are not directly involved in the new bond formation. It lowers the energy of the *endo*–transition state relative to that of the *exo*–transition state; hence the *endo*–adduct is obtained under kinetically controlled conditions. The *exo*–product is obtained at a high reaction temperature.

8. To monitor the reaction by TLC, co–spot the maleic anhydride and the product. As the maleic anhydride and the product are not visible under the UV lamp, use potassium permanganate ($KMnO_4$) dyeing reagent to check the formation of the product.
   <R_f of *cis*–Norbornene–5,6–*endo*–dicarboxylic anhydride is 0.90.>
   As the reaction proceeds, a white solid is formed.

9. The percent yield is 80~85 %.
   The purity of the resulting product is good enough for 200 MHz NMR and for the melting point without any further purification.

   m.p. of *cis*–Norbornene–5,6–*endo*–dicarboxylic anhydride: 159.0~161.0℃ (lit. value : 165~167℃)
   (for comparison ; m.p. of *exo* isomer : 140~145℃)

   ^{1}H NMR ($CDCl_3$) : δ 1.60 (1H), 1.80 (1H), 3.52 (2H), 3.58 (2H), 6.32 (2H)

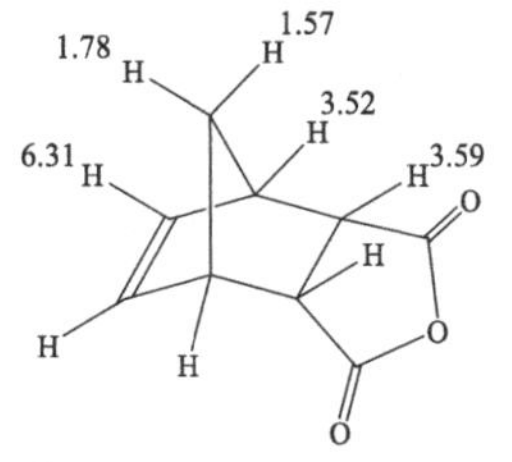

< ^{1}H Chmecal shifts of *cis*-norbornene -5,6-endo-dicarboxylic anhydride >

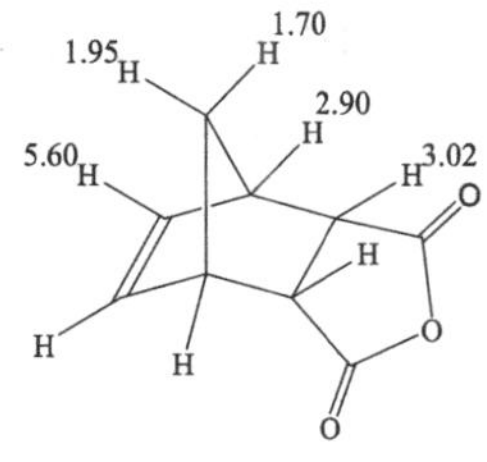

< Prediction of ^{1}H chemical shifts by Chem Draw >

10. The ***endo*–adduct** and the ***exo*–adduct** can be distinguished by the comparison of the **coupling constant** between the proton ***a*** and the *exo* or *endo* proton in each adduct. And it also provides valuable information about the carbon skeleton of the **[2.2.1] bicyclic system.**

    The coupling constant between the proton ***a*** and the *exo* proton in the *endo*–adduct is 2.8 Hz. It means that the **dihedral angle** between them is about 60°. It also confirms the *endo* structure assignment.

    In contrast, the coupling constant between the proton ***a*** and the *endo* proton in the *exo*–adduct is close to zero. Therefore, the dihedral angle between them is close to 90°. It also confirms the *exo* structure assignment.

    The coupling constant (close to zero) and the dihedral angle (90°) between the proton ***a*** and the *endo* proton are important characteristics of the carbon skeleton of the [2.2.1] bicyclic system.

<NMR interprestion of *endo* and *exo* isomer>

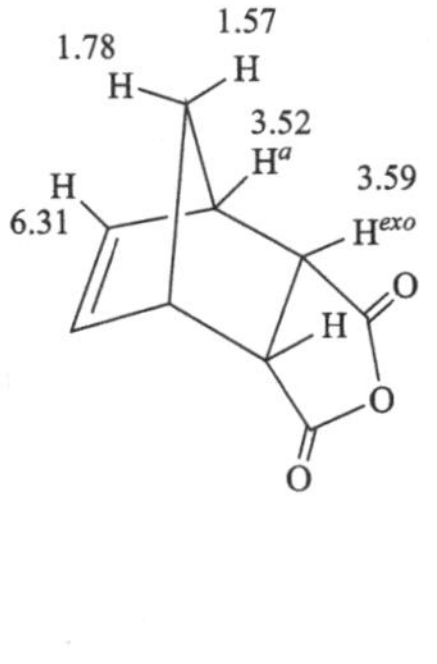

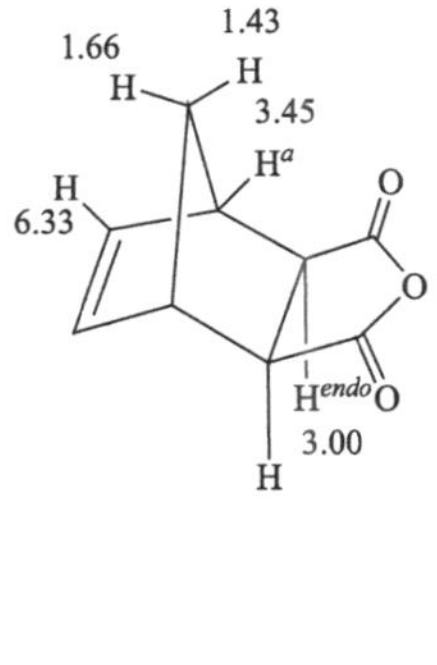

Keywords for Diels–Alder reaction

Chemical equation for Synthesis of cis–Norbornene–5,6–*endo*–dicarboxylic anhydride

Reaction Mechanism for Synthesis of cis–Norbornene–5,6–*endo*–dicarboxylic anhydride

CHAPTER

04

Appendix

4.1 Table of Isomers

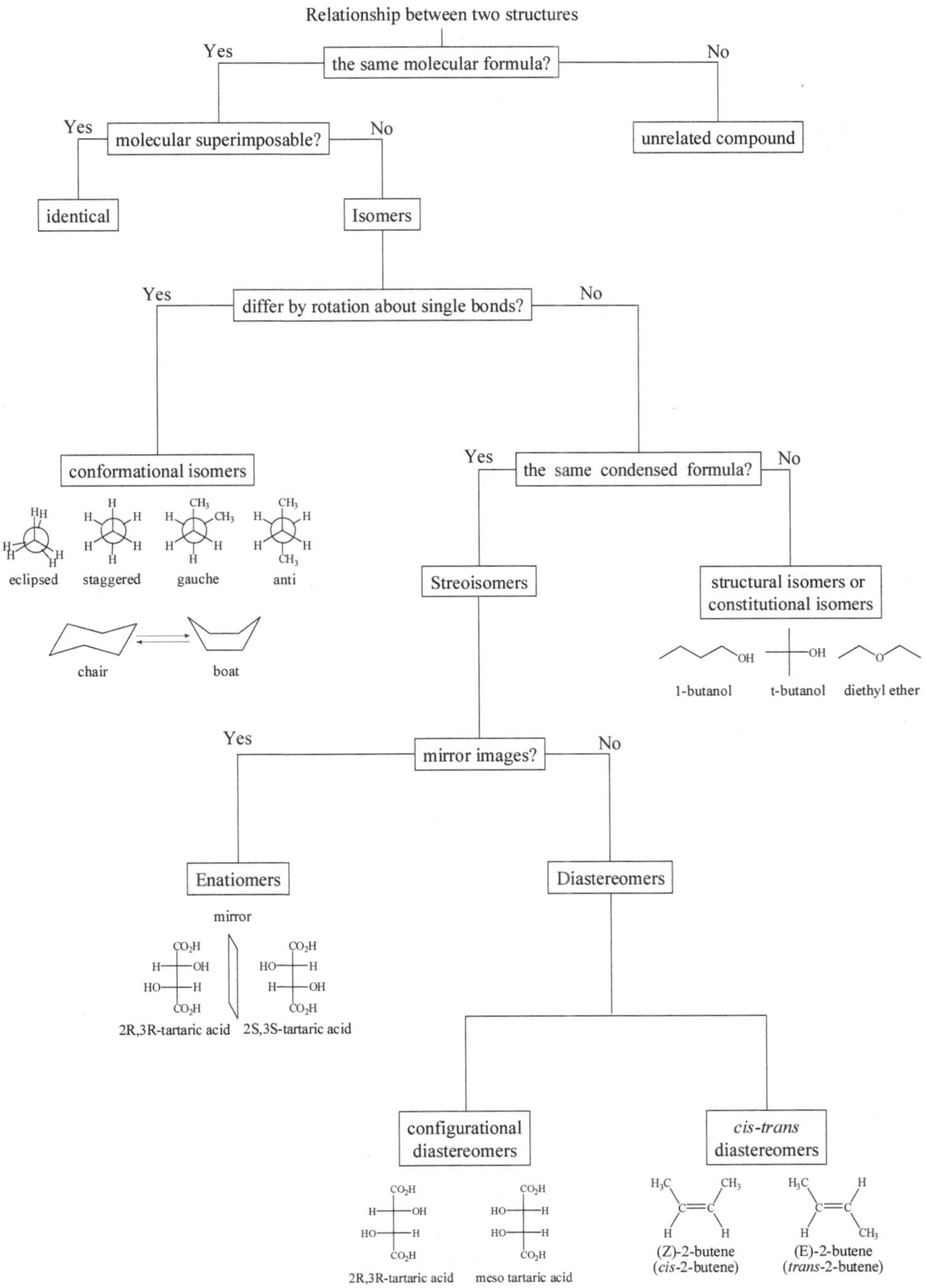

⟨Practice⟩

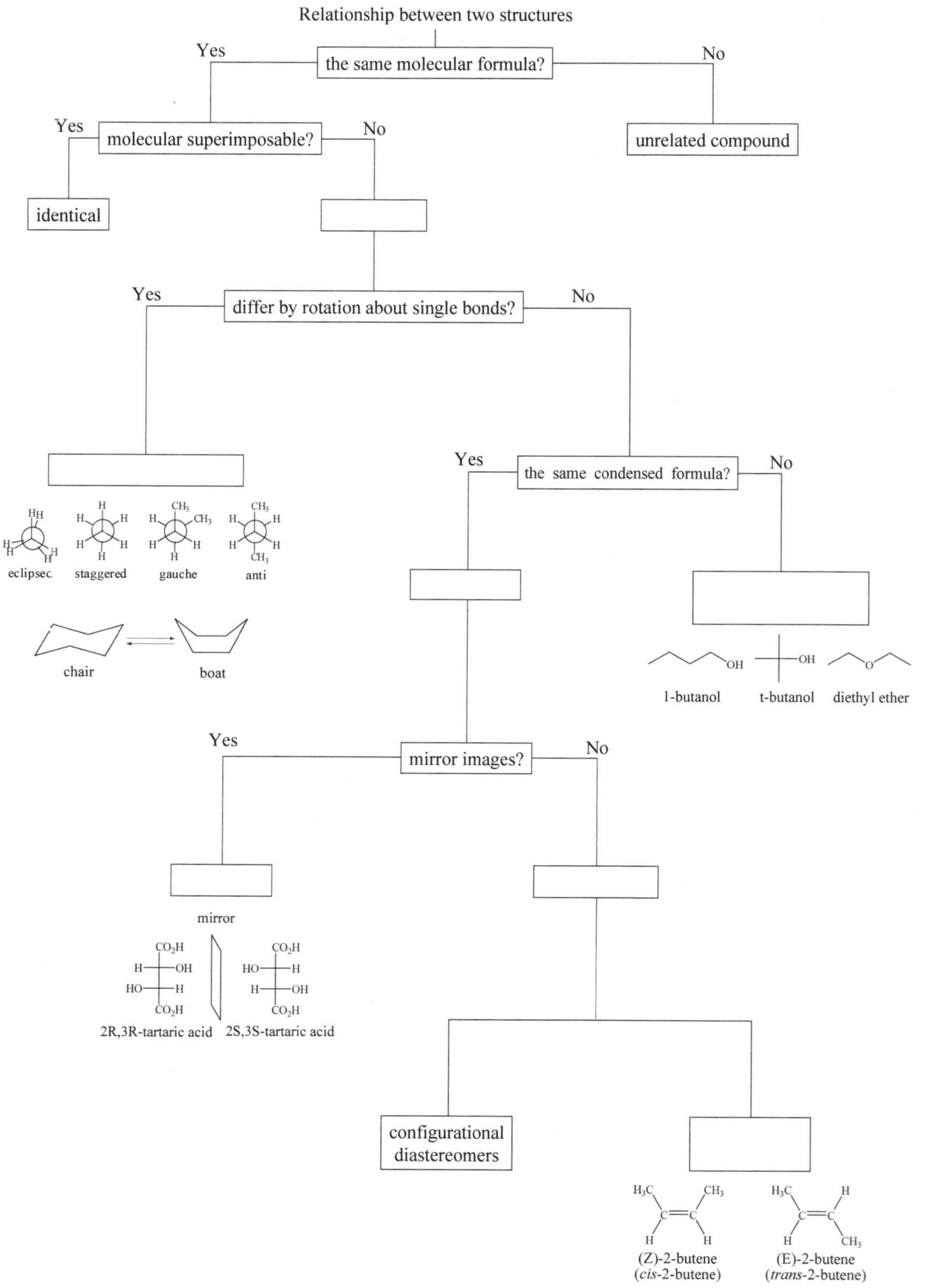
Relationship between two structures
the same molecular formula?
Yes
No
unrelated compound
molecular superimposable?
Yes
No
identical
differ by rotation about single bonds?
Yes
No
eclipsed
staggered
gauche
anti
chair
boat
the same condensed formula?
Yes
No
1-butanol
t-butanol
diethyl ether
mirror images?
Yes
No
mirror
2R,3R-tartaric acid
2S,3S-tartaric acid
configurational diastereomers
(Z)-2-butene
(cis-2-butene)
(E)-2-butene
(trans-2-butene)

4.2 Glassware and Apparatus

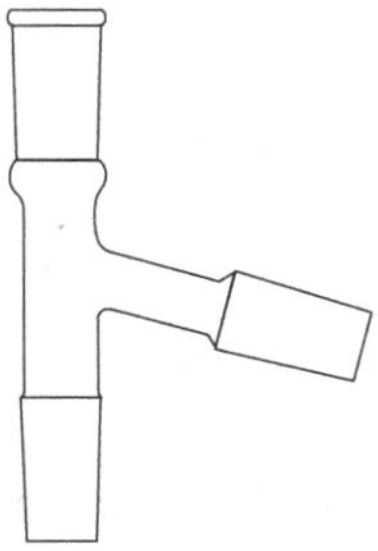

Distillation head

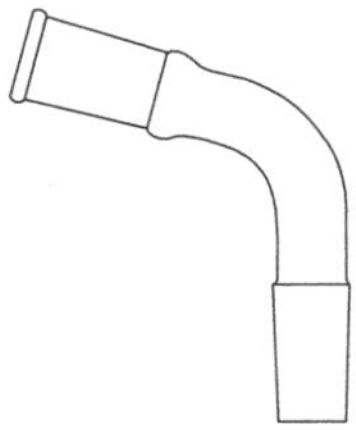

Adapter

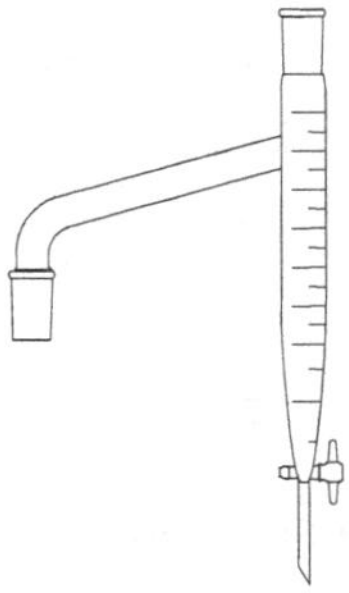

Dean-Stark

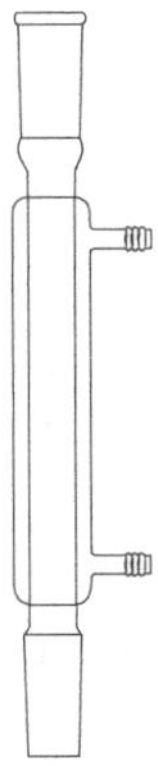

Liebig condenser

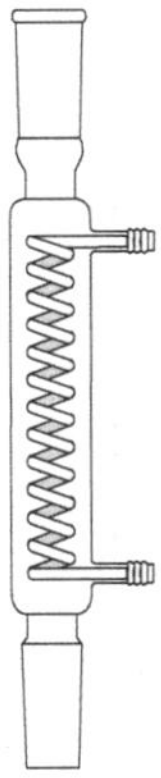

Reflux condenser

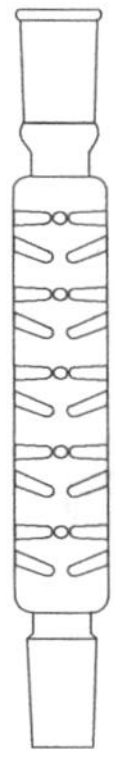

Vigreux condenser

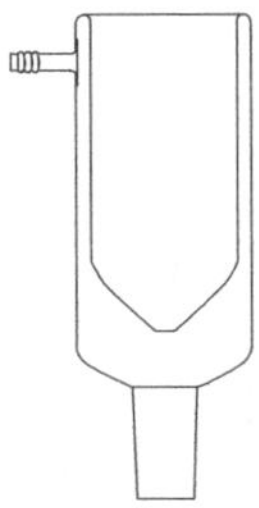

Dewar condenser

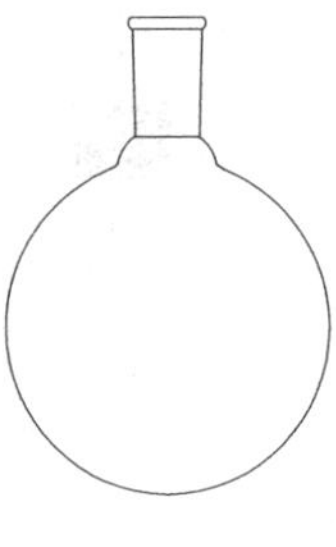

Round bottom flask

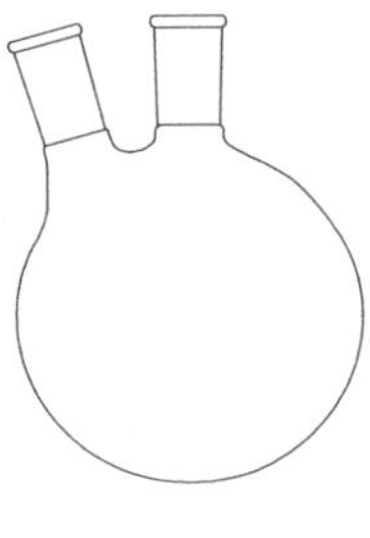

Two-neck flask

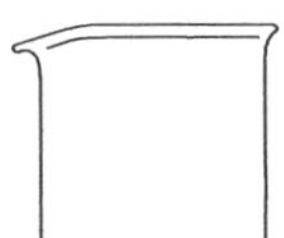

Beaker

Erlenmeyer flask

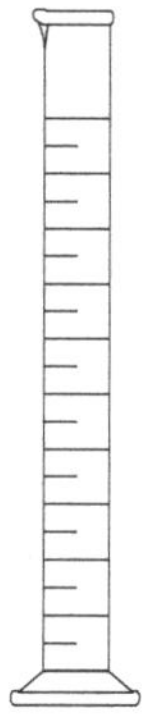
Graduated cylinder

Separatory funnel

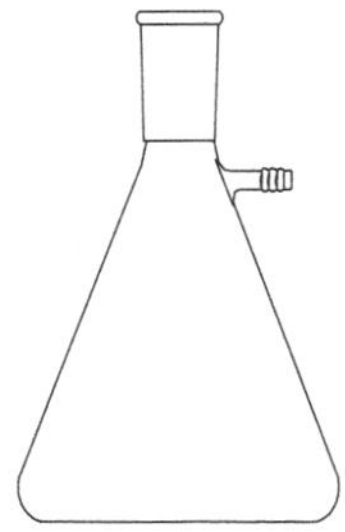
Filter flask

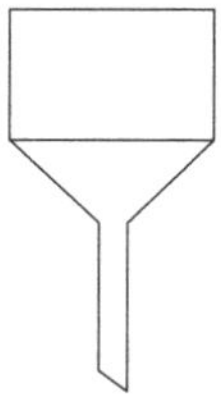
Büchner funnel

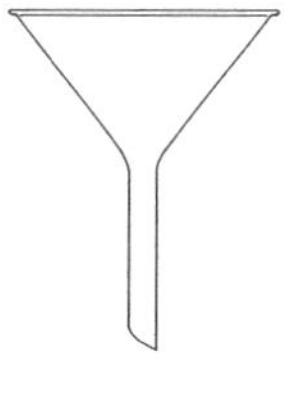
funnel

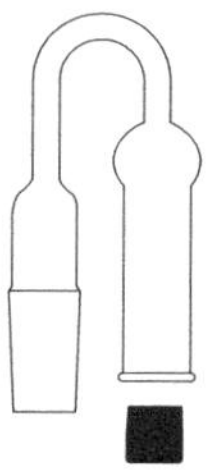
Drying tube adapter

Pipette

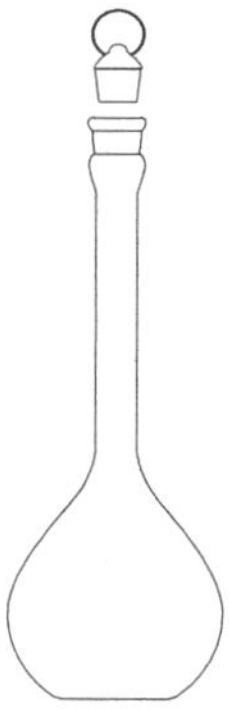
Volumetric flask

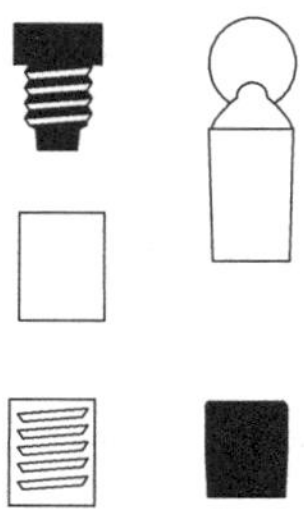
Stopper

Washing bottle

Thermometer

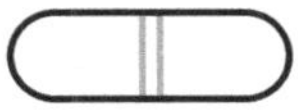

Magnetic spin-bar

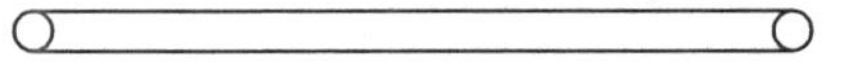

Glass rod

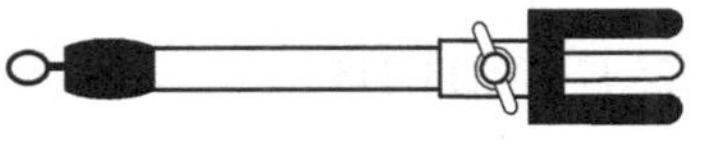

Utility clamp

Ring clamp

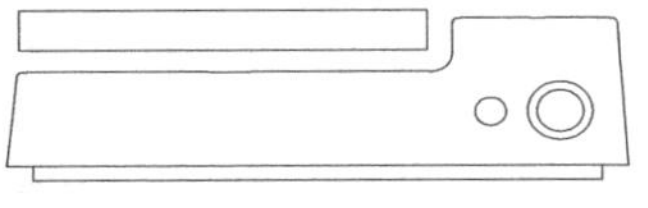

Magnetic stirrer

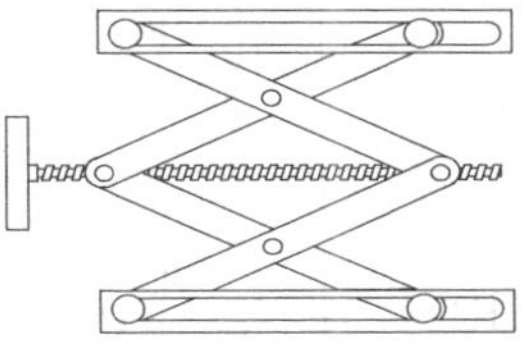

Lab jack

Heating mentle

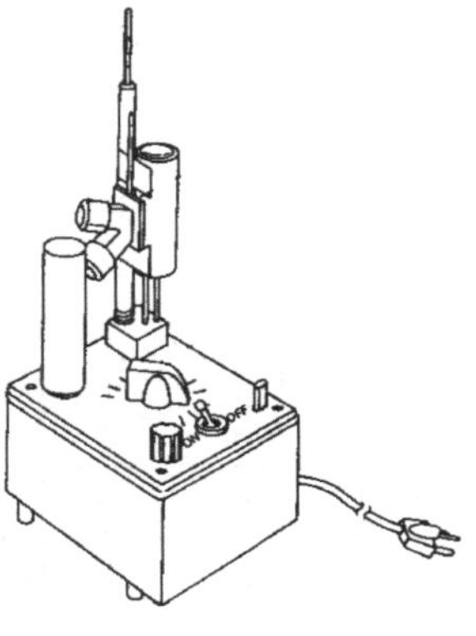

Melting point apparatus

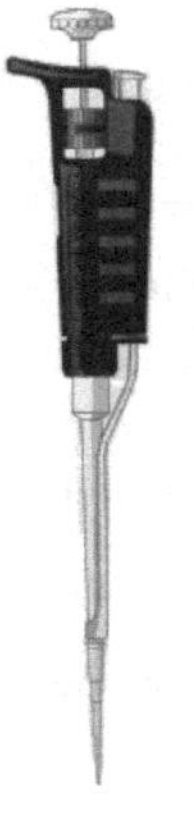

Micro pipette

4.3 Reagent

Reagent	Molecular formula	Structural formula	M.W.	m.p. (℃)	b.p. (℃)	density at 20℃ (g/mL)
Acetanilide	$C_6H_5NHCOCH_3$	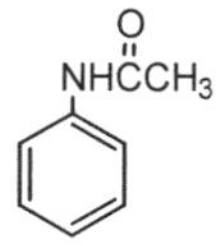	135.17	113~115		
Acetic acid	CH_3COOH		60.05		117~118	1.049
Acetic anhydride	$(CH_3CO)_2O$	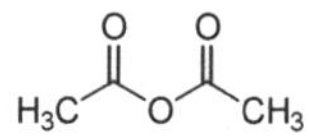	102.09	-73	138~140	1.082
Aniline	$C_6H_5NH_2$	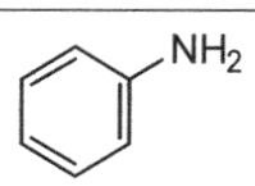	93.13	-6	184	1.022
Benzaldehyde	C_6H_5CHO	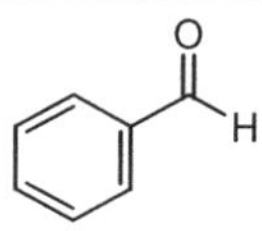	106.12	-26	178~179	1.044
Benzal bromide	$C_6H_5CHBr_2$	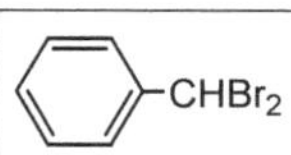	249.94		156	1.510
Benzoic acid	$C_6H_5CO_2H$	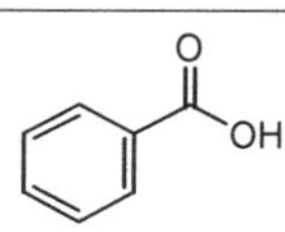	122.12	122~123	249.2	
Benzonitrile	C_6H_5CN	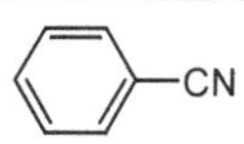	103.12	-13	191	1.010
Benzyl alcohol	$C_6H_5CH_2OH$	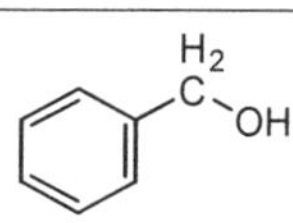	108.14	-15	205	1.045
Benzyl bromide	$C_6H_5CH_2Br$	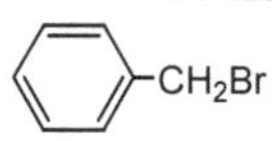	171.04	-3~-1	198~199	1.438
ε - Caprolactam	$C_6H_{10}NHO$	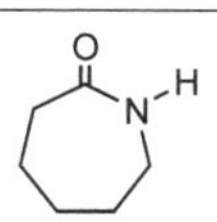	113.16	70~72	136~138	
Chloroform	$CHCl_3$		119.38	-63	61	1.492
Cyclohexanol	$C_6H_{11}OH$	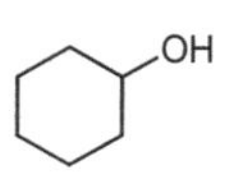	100.16	20~22	160~161	0.948
Cyclohexanone	$C_6H_{10}O$	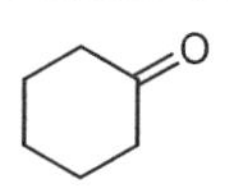	98.15	-47	155	0.947

Reagent	Molecular formula	Structural formula	M.W.	m.p. (℃)	b.p. (℃)	density at 20℃ (g/mL)
Cyclohexanone oxime	$C_6H_{10}NOH$	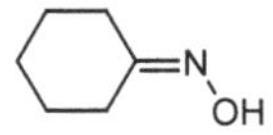	113.16	89~91	206~210	
Cyclopentadiene	C_5H_6		66.10	- 85	42	0.791
trans, *trans*-Dibenzylidene acetone	$C_6H_5CH{=}CHCOCH{=}CHC_6H_5$	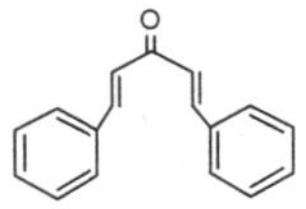	234.3	104~107		
Dicyclopentadiene	$C_{10}H_{12}$	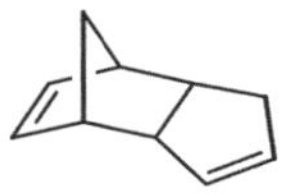	132.21	33	170	0.986
Diethylacetylene dicarboxylate	$C_8H_{10}O_4$	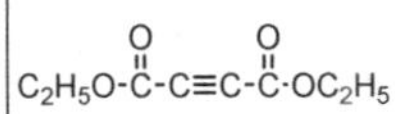	170.16		107~110	1.063
Diethyl fumarate	$C_8H_{12}O_4$	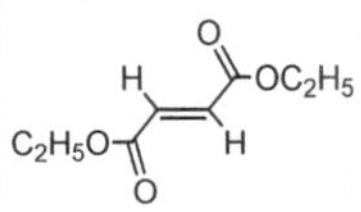	178.18	1~2	218~219	1.052
Diethyl maleate	$C_8H_{12}O_4$	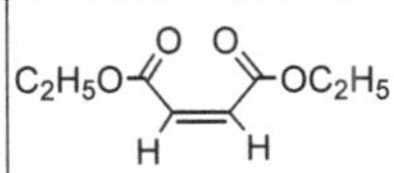	178.18	-8.8	225.3	1.069
N,N-Dimethylaniline	$C_6H_5N(CH_3)_2$	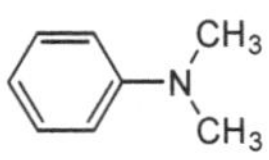	121.18		193~194	0.956
1,2-Dinitrobenzene	$C_6H_4(NO_2)_2$	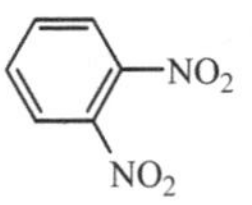	168.11	117~119		
1,3-Dinitrobenzene	$C_6H_4(NO_2)_2$	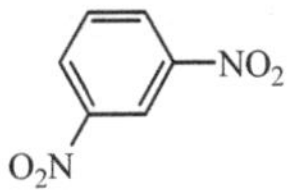	168.11	88~90	297	
1,4-Dinitrobenzene	$C_6H_4(NO_2)_2$		168.11	172~174	183.4 (34mmHg)	
2,3-Diphenyl-cyclopenta-2-enone	$(C_6H_5)_2C_5H_4O$	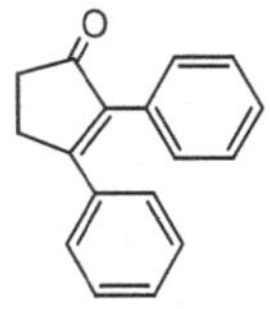	234.3			
Formic acid	HCO_2H		46.03	4℃ 이상	100~101	1.220
Hexylamine	$CH_3(CH_2)_5NH_2$		101.19	-23	131~132	0.766

Reagent	Molecular formula	Structural formula	M.W.	m.p. (℃)	b.p. (℃)	density at 20℃ (g/mL)
Hydrochloric acid	HCl		36.46			1.200
Hydroxylamine hydrochloride	$H_2NOH{\cdot}HCl$		69.49	약 151		
Lauryl alcohol	$CH_3(CH_2)_{11}OH$		186.34	24	259	0.831
Maleic anhydride	$C_4H_2O_3$		98.06	54~56	200	
Methyl alcohol	CH_3OH		32.04	-98	64	0.790
Methyl benzoate	$C_6H_5CO_2CH_3$	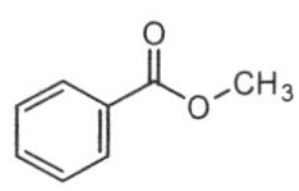	136.15	-12	198~199	1.094
Methyl *m* - nitrobenzoate	$O_2NC_6H_4CO_2CH_3$	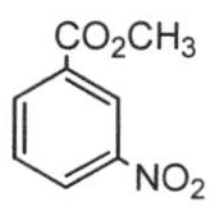	181.15	78~80	279	
Methyl orange	$(CH_3)NC_6H_4N=NC_6H_4SO_3Na$	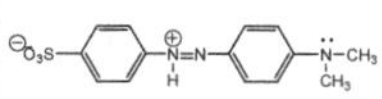	327.34			0.987
Methyl salicylate	$HOC_6H_4CO_2CH_3$	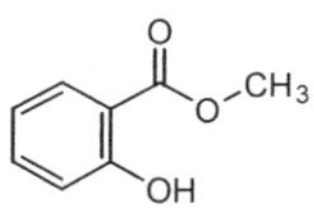	152.15	-7~-8	222	1.174
2-Naphthol	$C_{10}H_7OH$	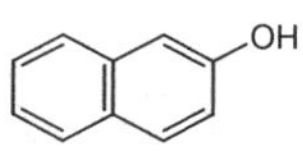	144.17	122~123	285~286	
Nitric acid	HNO_3		63.01	-42	86	1.490
o - Nitroaniline	$C_6H_4NH_2NO_2$	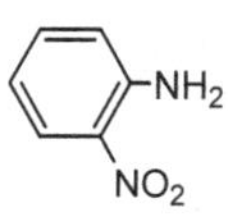	138.13	73~76	284	
p - Nitroaniline	$C_6H_4NH_2NO_2$	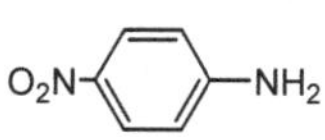	138.13	149~151	260	
Nitrobenzene	$C_6H_5NO_2$	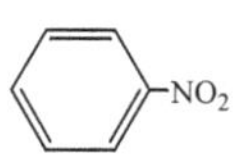	123.11	5~6	210~211	1.196

Reagent	Molecular formula	Structural formula	M.W.	m.p. (℃)	b.p. (℃)	density at 20℃ (g/mL)
o - Nitrobenzoic acid	$C_6H_4CO_2HNO_2$	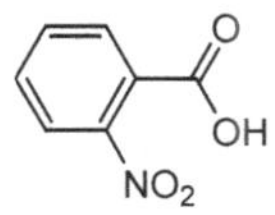	167.12	146~148		
p - Nitrobenzoic acid	$C_6H_4CO_2HNO_2$	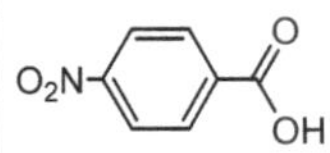	167.12	239~245		
o - Nitrophenol	$C_6H_4NO_2OH$	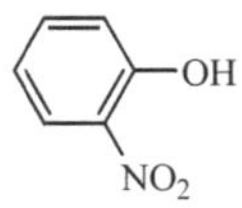	139.11	44~45	214~216	1.495
p - Nitrophenol	$C_6H_4NO_2OH$	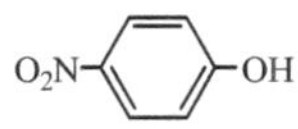	139.11	112~115		
cis-Norbornene-5,6-*endo*-2,3-dicarboxylic anhydride	$C_9H_8O_3$	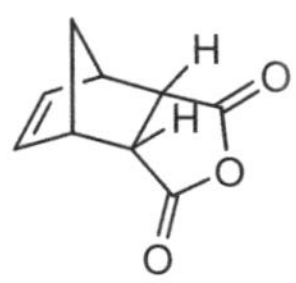	164.16	165~167		
Octylamine	$CH_3(CH_2)_7NH_2$		129.25	-5~-1	175~177	0.782
Orange Ⅱ	$HOC_{10}H_6N{=}NC_6H_4SO_3Na$	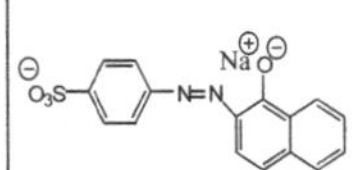	350.33			
Oxalic acid	HO_2CCO_2H		90.04	190		
Phenol	C_6H_5OH	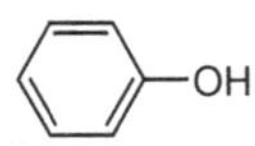	94.11	40~42	182	1.071
Phosphoric acid	H_3PO_4		98.00	41~44		1.830
Picric acid	$(O_2N)_3C_6H_2OH$	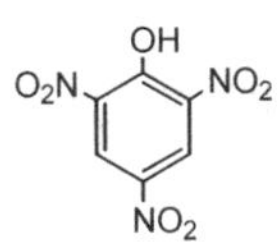	229.11	121~123		
Potassium hydroxide	KOH		56.11	361		
Pyridinium chlorochromate	$C_5H_5NHCrO_3Cl$	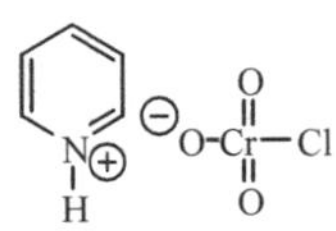	215.56	205		

Reagent	Molecular formula	Structural formula	M..W.	m.p. (℃)	b.p. (℃)	density at 20℃ (g/mL)
Salicylic acid	$HOC_6H_4CO_2H$	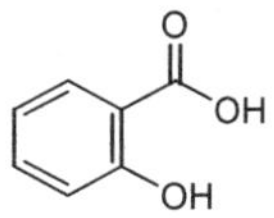	138.12	158~161	211	
Sodium acetate	CH_3COONa		82.03	328		1.528
Sodium borohyride	$NaBH_4$		37.83	400		1.074
Sodium carbonate	Na_2CO_3		105.99	851		
Sodium chloride	NaCl		58.44	801		2.165
Sodium dichromate dihydrate	$Na_2Cr_2O_7 \cdot 2H_2O$		298	91		2.350
Sodium hydroxide	NaOH		40	318		2.130
Sodium nitrite	$NaNO_2$		69	271		2.168
cis - Stilbene	$C_6H_5CHCHC_6H_5$	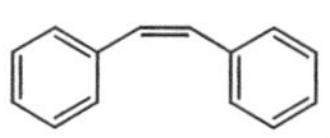	180.25		82~84	1.011
trans - Stilbene	$C_6H_5CHCHC_6H_5$	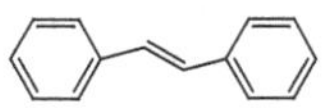	180.25	122~124	305~307	0.970
Sulfanilic acid	$H_2NC_6H_4SO_3H$	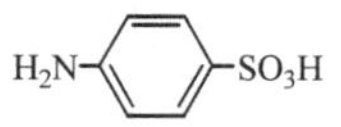	173.19	~288		
Sulfuric acid	H_2SO_4		98.08	-2	327	1.840
Tin	Sn		118.69	7.310		

4.4 Solvent

Solvent	b.p.(℃) (at 760 mmHg)	m.p. (℃)	M.W.	Density at 20℃ (g/mL)	Dielectric constants
Acetic acid	118	17	60	1.05	6.2
Acetone	56	-95	58	0.79	20.7
Acetonitrile	82	-44	41	0.78	37.5
Anisole	154	-38	108	0.99	4.3
1-Butanol	117.7	-90	74	0.81	17.1
2-Butanol	99.5	-115	74	0.81	15.8
tert-Butanol	82.4	26	74	0.79	10.9
Carbon disulphide	46	-111	76	1.26	2.6
Carbon tetrachloride	77	-23	154	1.59	2.2
Chloroform	61	-64	119	1.49	4.8
Cyclohexane	81	6.5	84	0.78	2.0
Dichloromethane	40	-95	85	1.33	8.9
Diethyl ether	35	-116	74	0.71	4.3
Dimethyl formamide	153	-60	73	0.95	36.7
Dimethyl sulfoxide	189	18	78	1.10	46.7
Ethyl acetate	77	-84	88	0.90	6.0
Ethyl alcohol	78	-114	46	0.79	24.6
Ethylene glycol	197	-16 ~ -13	62	1.11	37.7

Solvent	b.p.(℃) (at 760 mmHg)	m.p. (℃)	M.W.	Density at 20℃ (g/mL)	Dielectric constants
Ethyleneglycol dimethylether	83	-58	90	0.86	7.2
Furan	31.4	-86	68	0.95	2.9
Glycerol	290	18	92	1.26	42.5
Heptane	98	-91	100	0.68	9.1
Hexomethylphos-phoric triamide	233	7	179	1.03	30.0
Hexane	69	-95	86	0.66	1.9
Methyl alcohol	65	-98	32	0.79	32.0
Nitromethane	101	-29	61	1.14	35.9
Pentane	36	-130	72	0.63	1.8
Pyridine	115	-42	79	0.98	12.4
Tetrahydrofuran	66	-109	72	0.89	7.6
Toluene	111	-95	92	0.87	2.4

4.5 pK_a

구조	pK_a
$\underline{H}I$	-9
$\underline{H}Br$	-8
$\underline{H}Cl$	-7
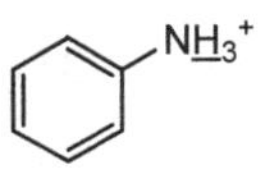	-10 ~ +5
$CCl_3COO\underline{H}$	0.5
$CHCl_2COO\underline{H}$	1.3
$HC{\equiv}CCOO\underline{H}$	1.9
$CH_2ClCOO\underline{H}$	2.8
$\underline{H}F$	3.2
$HCOO\underline{H}$	3.7
$H_2C{=}CHCOO\underline{H}$	4.2

구조	pK_a
$CH_3COO\underline{H}$	4.8
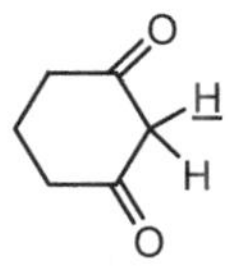	5.3
$HCO_3\underline{H}$	7.1
$CH_3CO_3\underline{H}$	8.2
$CH_3COC\underline{H}_2COCH_3$	9.0
$\underline{H}CN$	9.2
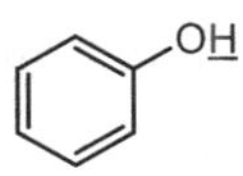	9.9
$RN\underline{H}_3^+$	10 ~ 12
$C\underline{H}_3NO_2$	10
$CH_3S\underline{H}$	10

구조	pK$_a$	구조	pK$_a$
$CH_3COC\underline{H}_2COOCH_3$	11	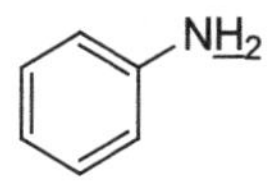	18~28
$C\underline{H}_2(COOCH_3)_2$	13	$HC{\equiv}C\underline{H}$	25
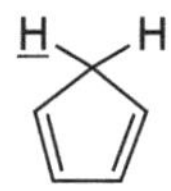	15	$C\underline{H}_3CN$	25
$CH_3O\underline{H}$	16	$CH_3SOC\underline{H}_3$	35
$\underline{H}_2O$	16	$\underline{H}_2$	35
$(CH_3)_3CO\underline{H}$	18	$N\underline{H}_3$	36
$RC\underline{H}_2CO_2R'$	19 ~ 20	$(CH_3CH_2)_2N\underline{H}$	40
$CH_3COC\underline{H}_3$	19	$H_2C{=}C\underline{H}_2$	44
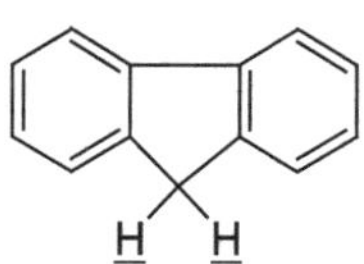	23	$C\underline{H}_4$	~60

4.6 Calculation of Concentration in Solution

Reagent	HCl	H_2SO_4	HNO_3
Molecular Weight	36.5	98.1	63.0
Weight % of conc. solution	37.3	96.5	70.0
Density of conc. solution (g/mL)	1.18	1.84	1.41
Molarity of conc. solution (M)	12.0	18.1	15.7
Weight % of dil. solution	20.0	25.0	32.0
Density of dil. solution (g/mL)	1.10	1.18	1.19
Molarity of of dil. solution (M)	6.03	3.01	6.04
Volume of conc. solution required for 1L of 1M solution (mL)	82.6	55.2	63.7

* Weight % of conc. solution is varied slightly depending on the suppliers. Read the label carefully.

1. Calculating the molarity (M) of the concentrated hydrochloric acid solution (conc. HCl)

Molarity means the number of moles of solute per 1.00 L of the solution. The hydrochloric acid solution which has 37.3 wt % of mass density contains 37.3 g of acid per 100g of solution. By using the hydrochloric acid solution's density 1.18 g/mL and its molecular weight 36.5 g/mol, the wt % can be converted into molarity.

① The volume of solution can be calculated by dividing the 100 g of solution with its density.

$$100\text{ g} \times \frac{1.0\text{ mL}}{1.18\text{ g}} \times \frac{1.00\text{ L}}{1000\text{ mL}} = 0.0847\text{ L}\ (= 84.7\text{ mL})$$

② In 100 g of solution, it contains 37.3 g of acid, so find the number of mol in 37.3 g of acid.

$$37.3\ g \times \frac{1\ mol}{36.5\ g} = 1.02\ mol$$

③ Convert the number of mol in ② into the volume of 1.00 L of the solution by its ratio equation.

$$X\ mol : 1.00\ L = 1.02\ mol : 0.0847\ L$$

$$X\ mol = \frac{1.02\ mol \times 1\ L}{0.0847\ L} = 12.0\ mol$$

④ 12.0 moles of the acid is dissolved in the 1.00 L of the solution, so the molarity of conc. HCl is 12.0 M.

2. How to prepare a dilute solution

① **The number of mole in a solute does not change even if the amount of the solvent changes.** By using this principle, the dilute solution can be easily made from the concentrated hydrochloric acid by formula.

$$M \times V = M' \times V' = X\ mole$$

ex) Preparing 1.00 L of 3.00 M HCl solution with concentrated hydrochloric acid

$$3.00\ M \times 1.00\ L = 12.0\ M \times Y\ L$$
$$Y = 0.250\ L\ (= 250\ mL)$$

1) Fill a half of the 1.00 L volumetric flask **A** with the distilled water.
2) Transfer 250 mL of conc. HCl on the flask **A** and stir well.
3) Fill the distilled water to the marked line.

② Making dilute solution with a solid sample
From the concentration and amount of the desired solution, calculate the moles of the solute.

$$M' \times V' = X\ mole$$
$$X\ mole \times Molecular\ Weight = weight\ of\ a\ solid\ sample$$

ex) Making 1.00 L of 3.00 M NaOH solution

$$\frac{3\ \text{mol}}{1\ \text{L}} \times 1\ \text{L} \times \frac{40.0\ \text{g}}{1\ \text{mol}} = 120\ \text{g}$$

1) Add the distilled water in 500 mL beaker **B** in an ice bath, and dissolved 120 g of NaOH slowly.
2) Transfer the solution in **B** to the 1.00 L volumetric flask **C**. Wash the beaker **B** with a small amount of distilled water and add it to 1.00 L volumetric flask C. Fill it with the distilled water to the marked line.

3. The weight % solution

ex) Preparation of 20.0 wt % NaCl solution

Make the mass of solute to be a 20 % of the resulting solution.

$$20\ \% = \frac{x\ \text{g (NaCl)}}{100\ \text{g } (H_2O) + x\ \text{g (NaCl)}} \times 100$$

x = 25 g (When using 100 mL of the solvent)

∴ Dissolve 25.0 g of NaCl in 100 mL of the distilled water.

4. The volume % solution

ex) Preparation of 20.0 vol % ethanol solution

Make the volume of solute to become 20.0 % of the resulting solution.

$$20\ \% = \frac{x\ \text{mL (ethanol)}}{100\ \text{mL } (H_2O) + x\ \text{mL (ethanol)}} \times 100$$

x = 25.0 mL (When using 100mL of water as solvent)

∴ Add 25 mL of ethanol in 100 mL of water.

Caution : When the volume of the resulting solution is smaller than the total of volume of the solute and the solvent, the equation above should be calibrated. Therefore, the molarity and wt

% are recommended in the laboratory.

5. The molarity of the common solution

It is assumed that the density of the water is 1.0 g/mL, and the solute does not give any effect to the volume of the solution.

① Saturated sodium chloride solution (brine)
Molecular weight: 58.4 g/mol
Solubility: 36.0 g / 100.0 mL
36.0 g (NaCl) ÷ 58.4 g/mol (NaCl) ÷ 0.100 L = 6.16 M (mole/L)

② Saturated $NaHCO_3$ solution
Molecular weight: 84.0 g/mol
Solubility: 8.00 g / 100.0 mL
8.00 g ($NaHCO_3$) ÷ 84.0 g/mol ($NaHCO_3$) ÷ 0.100 L = 0.952 M

③ 20 wt % of $NaHSO_3$ solution
Molecular weight: 104.1 g/mol
Solubility: 29.0 g / 100. mL
20.0 g ($NaHSO_3$) ÷ 104.1 g/mol ($NaHSO_3$) ÷ (0.100 – 0.020) L = 2.40 M

④ 10 wt % of NaOH solution
Molecular weight: 40.0 g/mol
Solubility: 108.0 g / 100. mL
10.0 g (NaOH) ÷ 40.0 g/mol (NaOH) ÷ (0.100 – 0.010) L = 2.78 M

4.7 Dyeing Reagent for Thin Layer Chromatography

The preparations of the dyeing reagents are the following.

1. Sulfuric acid reagent

Dissolve 10.0 mL of sulfuric acid (H_2SO_4) in 85.0 mL of the distilled water and then add 5.0 mL of methanol.

2. Potassium permanganate reagent

Dissolve 1.50 g of potassium permanganate ($KMnO_4$) in 300.0 mL of the distilled water. Add 10.0g of potassium carbonate (K_2CO_3) and 5.0 mL of 5 % sodium hydroxide (NaOH) to the resulting solution.

3. Vanillin – phosphoric acid reagent

Dissolve 1.00 g of vanillin in 25.0 mL of ethanol. Then add 25.0 mL of the distilled water and 35.0 mL of phosphoric acid (H_3PO_4) to the resulting solution.

4. Anisaldehyde – sulfuric acid reagent

Add 2.0 mL of sulfuric acid (H_2SO_4) to 1.0 mL of anisaldehyde, and then add 100.0 mL of acetic acid.

4.8 A Simple Calculation of ^1H NMR Shifts

1. The chemical shift of a proton is the sum of the following values based on electronegativity.

H	B	C	N	O	F
0.2	-0.3	0.7	1.7	2.7	3.7
		Si	P	S	Cl, Br, I
		-0.3	0.7	1.7	2.7

In addition, there are values for the deshielding of the unsaturated bonds caused by ring–current effect (diamagnetic anisotropy).

$H_2C{=}CH$–**H** vinyl (Vn)	4.2	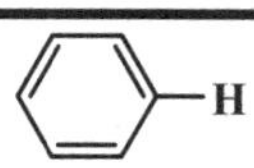 aryl (Ar)	6.0
$H_2C{=}CH{-}CH_2$–**H** allyl (Al)	0.8	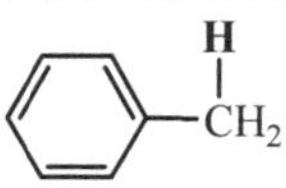 benzyl (Bz)	1.2
$HC{\equiv}C$–**H** ethynyl (En)	2.1	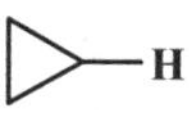 3-membered ring	- 0.7
$HC{\equiv}C{-}CH_2$–**H** propargyl (Pg)	0.8	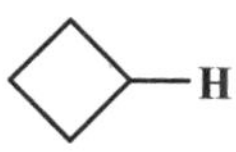 4-membered ring	0.7

Ex) ① Ethanol

$\overset{H^a}{CH_2}-\overset{H^b}{CH}-OH$

(a) $2H + C = 0.2 \times 2 + 0.7 = 1.1$ (observed at 1.22 *ppm*)

(b) $H + C + O = 0.2 + 0.7 + 2.7 = 3.6$ (observed at 3.70 *ppm*)

② Acetone

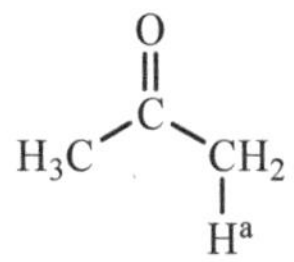

(a) 2H + C + Al = 0.2 × 2 + 0.7 + 0.8 = 1.9 (observed at 2.05 *ppm*)

③ Toluene

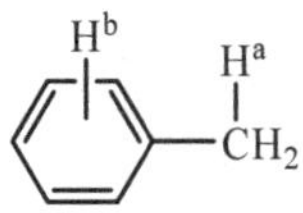

(a) 2H + C + Bz = 0.2 × 2 + 0.7 + 1.2 = 2.3 (observed at 2.32 *ppm*)
(b) 2C + Ar = 2 × 0.7 + 6.0 = 7.4 (observed at 7.17 ∓ 0.03 *ppm*)

④ Cyclopropane

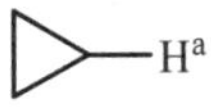

(a) H + 2C + 3–membered ring = 0.2 + 0.7 × 2 + (–0.7) = 0.9 (observed at 0.70 *ppm*)

⑤ Acetonitrile

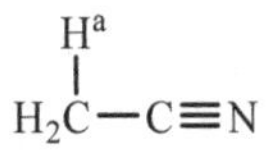

(a) 2H + C + Pg = 0.2 × 2 + 0.7 + 0.8 = 1.9 (observed at 1.95 *ppm*)

⑥ TMS

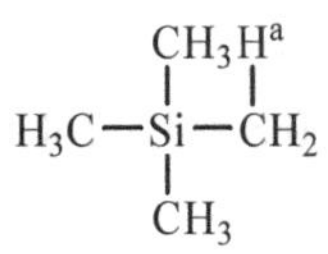

(a) 2H + Si = 0.2 × 2 + (–0.3) = 0.1 (observed at 0.0 *ppm*)

2. The allyl and benzyl effects are transmitted across from oxygen, or nitrogen atom to the proton on a saturated carbon atom.

Ex) ① Ethyl acetate

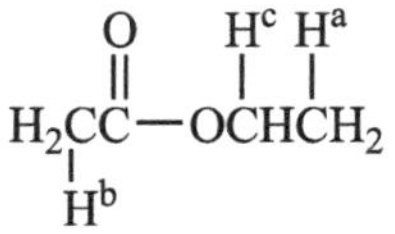

(a) 2H + C = 0.2 × 2 + 0.7 = 1.1 (observed at 1.25 *ppm*)

(b) 2H + C + Al = 0.2 × 2 + 0.7 + 0.8 = 1.9 (observed at 2.03 *ppm*)

(c) H + C + O + Al = 0.2 + 0.7 + 2.7 + 0.8 = 4.4 (observed at 4.12 *ppm*)

② DMF

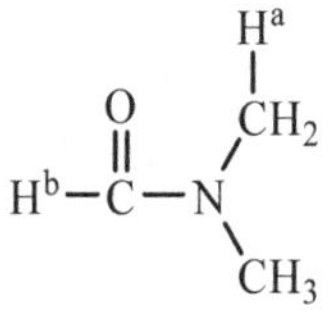

(a) 2H + N + Al = 0.2 × 2 + 1.7 + 0.8 = 2.9 (observed at 2.88, 2.97 *ppm*)

(b) N + O + Vn = 1.7 + 2.7 + 4.2 = 8.6 (observed at 8.02 *ppm*)

3. The cumulative effect of oxygen and fluorine atom is diminished. For example, the first oxygen atom has its full value–2.7– but the second oxygen holds 1.7. The value of fluorine goes from 3.7 to 2.7–

Ex) 4,6–dioxacycloheptene

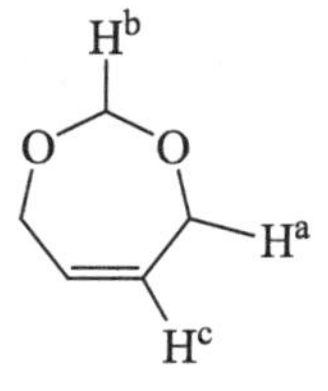

(a) H + C + O + Al = 0.2 + 0.7 + 2.7 + 0.8 = 4.4 (observed at 4.30 *ppm*)

(b) H + O + 2nd O = 0.2 + 2.7 + 1.7 = 4.6 (observed at 4.90 *ppm*)

(c) 2C + Vn = 0.7 × 2 + 4.2 = 5.6 (observed at 5.70 *ppm*)

For more detail, Refer to Jeffreys, J. A. D. *Journal of Chemical Education* **1979**, 56, 807

Index

B

C

D

E

F

G

H

I

N

Q

R

S

K's Manual for Organic Laboratory

| Author Hongbum Kim
| Publisher Jiyoung Kim

| Address 37-42, Hoedong-gil, Paju-si, Gyeonggi-do, Korea
| Tel 031-955-1321
| Fax 031-955-1322
| Web Site www.freeaca.com
| E-mail main@freeaca.com, editor@freeaca.com(editor), crm@freeaca.com
| Registration 406-2003-017, 1980. 7. 12.
| Publication 15, December, 2006.
| Price 17,000 won

Printed in Korea

ISBN 89-7338-603-4 93430